AF601245

# HRM in Agri-Business

## About the Author

**Dr V M Thumar** obtained his B. Sc. (Agriculture) & M. Sc. (Agricultural Economics) from Gujarat State Agriculture University, Gujarat. He completed his Ph.D. (Economics) degree from Junagadh Agricultural University, Junagadh, Gujarat. Dr. Thumar presently working as Associate Professor in ASPEE College of Agriculture, Navsari Agricultural University, Navsari. Besides this Dr. Thumar is also serving as Planning Officer in Navsari Agricultural University, Navsari.

He has 34 years of experience in the area of Agricultural Economics, Human Resource Management, Financial Management, Export, Agriculture Marketing, Computer Science and Information Technology and presented many paper in seminars/conferences. He has participated in several workshops and training programmes. He has supervised several MBA and M. Sc. Students in Agricultural Economics.

# HRM in Agri-Business

Dr V M Thumar

2018
Daya Publishing House®
*A Division of*
Astral International Pvt. Ltd.
New Delhi – 110 002

ISBN: 9789388173032 (Int. Edition)



*Published by* : **Daya Publishing House®**
*A Division of*
**Astral International Pvt. Ltd.**
– ISO 9001:2015 Certified Company –
4736/23, Ansari Road, Darya Ganj
New Delhi-110 002
Ph. 011-43549197, 23278134
E-mail: info@astralint.com
Website: www.astralint.com

*Digitally Printed at* : **Replika Press Pvt. Ltd.**

# Preface

The focus of HRM is on managing people within the employer-employee relationship. It involves the productive use of people in achieving the organisation's strategic business objectives and the satisfaction of individual employee needs. The 'human resources' of an organisation consist of the workers who perform its task activities, functions in exchange for wages, salaries and benefits. HRM is a part of management dealing directly with people, whereas management also includes marketing, operations management, research & development, accounting and finance.

Agri-business as a concept was born in Harvard University in 1957 with the publication of a book "A concept of Agri-business", written by John David and A. Gold Berg. It was introduced in Philippines in early 1966, when the University of the Philippines offered an Agri-business Management (ABM) programme at the under-graduate level. In 1969, the first advanced Agri-business management seminar was held in Manila. "Agri-business is the sum total of all operations involved in the manufacture and distribution of farm supplies, production activities on the farm, storage, processing and distribution of farm commodities and items made from them".

Agri-business is sum total of all operations or activities involved in the business of production and marketing of farm supplies and farm products for achieving the targeted objectivities. It deals with agricultural sector and also with the portion of industrial sector, which is the major source of farm inputs like fertilizers, pesticides, machines, processing and postharvest technologies.

There might have been few Human Resource Management is the process of proper and maximise utilisation of available limited skilled workforce. The core purpose of the HRM is to make efficient use of existing human resources in the organisation. ABM applied business theories and practices to the agricultural industry for lower costs boost profits and ensure that farm or food products are grown and distributed effectively.

The present book entitled *"**HRM in Agri-Business**"* comprises of eleven chapters. Efforts have been made to incorporate information on introduction to human resources management, introduction to human resource planning, human resources and business growth, career planning, training and development, agri-business planning, HRM in Agri-business expansion, agri-business financial management, working capital management in agri-business, agri-business capital budgeting compensation management and HRM training. Detailed glossary of important terms are provided at the end of each chapter will enable the readers to update their knowledge.

It is hoped that the book is quite useful for students, teachers and researchers of management, commerce, economics, agri-business management, policy makers and people having concern for HRM and agri-business management errors in-spite of best efforts made through careful editing and reading. Therefore, readers are sincerely requested to make suggestions for the improvement of this publication to augment its full ability and wider acceptability.

# Contents


*Preface* *v*

**1. Introduction to Human Resource Management** **1**

- 1.1 *Introduction*
- 1.2 *Definition*
- 1.3 *Concept*
- 1.4 *Importance*
- 1.6 *Role of Human Resource Management*
- 1.7 *Functions of HRM*
- 1.8 *Personnel Management v/s Human Resource Management*
- 1.9 *Human Resource Organisation*
- 1.10 *Human Resource Management in India*
- 1.10 *Human Resource Development*
- *Questions*
- *Glossary*

**2. Introduction to Human Resource Planning** **17**

- 2.1 *Introduction*
- 2.2 *Human Resource Planning*
- 2.3 *Objectives of HR Planning*
- 2.4 *Importance of HR Planning*
- 2.5 *Benefits of HR Planning*
- 2.6 *Process of Human Resource Planning*
- 2.7 *Job Analysis*
- 2.8 *Job Description*

2.9 *Job Specification*
2.10 *Retaining Your Highly Skilled Staff*
2.11 *Managing an Effective Downsizing Programme*
2.12 *Strategic Human Resource Planning*
2.13 *Strategic Human Resource Planning Model*
2.14 *Planning the Total Workforce*
2.15 *Investing in Human Resource Development and Performance*
*Questions*
*Glossary*

**3. Human Resources and Business Growth** **41**

3.1 *Introduction*
3.2 *Human Resource in Small Business*
3.3 *Literature Review and Hypotheses Development*
3.4 *Recruitment and Selection*
3.5 *Training and Performance Appraisal*
3.6 *HRM in Business Growth*
3.7 *The Growth Gap*
3.8 *Four Roles of HRM*
3.9 *Human Resource Planning in Business*
3.10 *Outsourcing HR Reaps Benefits for Small Business*
3.11 *Changing Organisational Structures/Work Patterns in "SHAMROCK ORGANISATIONS"*
3.12 *Training in Small Industries*
*Questions*
*Glossary*

**4. Career Planning, Training and Development** **63**

4.1 *Introduction*
4.2 *Career Planning*
4.3 *Significance of Career Planning*
4.4 *Recruitment*
4.5 *Source of Recruitment Selection*
4.6 *Selection*
4.7 *Steps of Selection Procedure*
4.8 *Training and Development*
4.9 *Evaluation of Training and Development Programme*

*Questions*
*Glossary*
**5. Agri Business Planning** 83
5.1 *Introduction*
5.2 *Business Plan*
5.3 *SWOT Analysis*
*Questions*
*Glossary*
**6. HRM in Agribusiness Expansion** 97
6.1 *Introduction to Agribusiness*
6.2 *Sectors*
6.3 *Importance and Dimensions of Agri-business*
6.4 *HRM in Agriculture*
6.5 *Nature of HRM in Agri-business*
6.6 *Significance of HRM in Agri-business*
6.7 *HR in Agribusiness Expansion*
6.8 *Writing Agribusiness Plan*
*Questions*
**7. Agribusiness Financial Management** 115
7.1 *Agribusiness Financial Records*
7.2 *Agribusiness Financial Records*
7.3 *Types of Agribusiness Financial Statements*
7.4 *Financial Management*
*Questions*
**8. Working Capital Management in Agri-business** 139
8.1 *Introduction*
8.2 *Meaning and Definition*
8.3 *Types of Working Capital*
8.4 *Factors Determining Working Capital*
8.5 *Operating Working Capital Cycle*
8.6 *Working Capital Requirements*
8.7 *Estimating Working Capital Needs and Financing Current Assets*
8.8 *Inventory Management*
*Questions*
*Glossary*

**9. Agribusiness Capital Budgeting** 155

*9.1 Introduction*

*9.2 Meaning of Capital Budgeting*

*9.3 Principal of Capital Budgeting*

*9.4 Kinds of Capital Budgeting Proposals*

*9.5 Capital Budgeting Techniques*

*9.6 Estimation of Cash Flow for New Project*

*9.7 Sources of Long Term Funds*

*Questions*

*Glossary*

**10. Compensation Management** 163

*10.1 Introduction*

*10.2 Meaning*

*10.3 Today's Modern Compensation Systems*

*10.4 Components of Compensation System*

*10.5 Types of Compensation*

*10.6 Compensation Process*

*10.7 Techniques of Job Evaluation*

*Questions*

*Glossary*

**11. HRM Trainings** 179

*11.1 Human Resources*

*11.2 Developing an Identity for Your Agri Business*

*11.3 Recruiting*

*11.4 Hiring*

*11.5 Employee Orientation and Training*

*11.6 Communication*

*11.7 Motivate Employee Performance*

*11.8 Employee Manuals*

*11.9 Resolving Employee Conflict*

*11.10 Handling Discipline Issues Effectively*

*Glossary*

*References* **191**

# Unit 1

# Introduction to Human Resource Management

## 1.1 Introduction

Human Resource is considered the prime asset of an organization in modern times and their proper maintenance has become an important objective of the management. Human resource management is a growth oriented concept and considered an off shoot of traditional system of personnel management. The prime objective of Human resource management is therefore to integrate all the functions, resources and processes of an organization. It is concerned with the management and development of people of the organization.

In the current business scenario HRM has evolved to be an active partner in driving organizational growth and success. It aims at utilizing the potential of employees by providing those best inputs that fulfills their needs and desires. HRM is performance oriented and has many sub divisions like organizational development, performance management systems etc.

Human Resource Management is the study of activities about the workforce in an organization. It involves an administrative function that attempts to go with the organizational wants and the skills and abilities of its employees. Let us see what is meant by the three key terms: Human, resource and management.

- The word 'Human' means Homo-sapiens or Social Animal

- ☆ Resources mean assets such as Human, Physical, Financial, and Technical, Informational etc.
- ☆ Management is related to the function of Planning organizing, leading and controlling of organizational resources to accomplish goals efficiently and effectively

To put it simply, human resource management is responsible for how people are managed in the organizations. It is responsible for bringing people in an organization and helps them perform their work, compensating them for their work and solving problems that arise.

Human resource applies to the workforce managed by any employer. A business of any size needs employees in order for it to run. As an important – the most important – asset for any business leader, employees need to be properly managed, in order for optimal efficacy to be achieved.

## 1.2 Definition

"According to Michael J Jucious, Human resource management may be defined as that field of management which has to do with planning, organizing and controlling the functions of procuring, developing, maintaining and utilizing a labor force, such that the:

(a) Objectives for which the company is established are attained economically and effectively,

(b) Objectives of all levels of personnel are served to the highest possible degree,

(c) Objectives of society are duly considered and served".

**Fig. 1.1: Human Resource Management**

Another definition has been given by three eminent researchers Scott, Clothier and Spriegel who defined human resource management as "that branch of management which is responsible on a staff basis for concentrating on managing relationship between employers and employees, and with the development of the individual and the group. The objective is to attain maximum individual development, desirable working relationship between employers and employees and employees and employees and effective moulding of human resources as contrasted with physical resources".

"An important concept was established by Dale Yoder. According to him the management of human resources is viewed as a system in which participants seeks to attain both individual and group goals".

"According to Flippo, human resource management is the planning, organising, directing and controlling of the procurement, development, compensation, integration, maintenance and reproduction of human resources to the end that individual, organisational and societal objectives are accomplished".

"An important definition in this context has been given by National Institute of Personnel Management of India, Personnel management (or human resource management) is that part of management which is concerned with people at work and with their relationships within the organisation. It seeks to bring together men and women who make up an enterprise, enabling each to make his best contribution to its success both as an individual and as a member of a working group".

Human Resource Management (HRM) therefore is a function within an organisation that focuses on management of employees as well as providing direction to people who work in the organisation. It is the organisational function that deals with issues related to people such as compensation, hiring, performance management, safety, wellness, benefits, employee motivation, communication, administration, and training.

HRM is concerned with the management of human resources in an organisation. It tries to secure the best from people by winning their wholehearted cooperation. In short, it may be defined as the art of procuring, developing and maintaining competent workforce to achieve the goals of an organisation in an effective and efficient manner.

## 1.3 Concept

1. **People Oriented** - Human resource management is concerned with managing employees both as individuals and as group to achieve organizational goals. It is also concerned with managing behavior, emotions and social aspects of personnel. It is the process of bringing people and organizations together in order to achieve their respective goals.
2. **Individual Oriented** - Under human resource management, every employee is considered as an individual. The services and programs are designed in a way that the needs of each employee can be fulfilled thereby leading to employee satisfaction and growth. In other words, it is concerned with the development of human resources, *i.e.* knowledge, capability, skill, potentialities and achieving individual goals.
3. **Continuous Function** - Human resource management is a continuous and never-ending process. According to George R Terry, "it cannot be

turned on and off like water from a faucet; it cannot be practiced only one hour each day or one day each week. Personnel management requires a constant alertness and awareness of human relations and their importance in everyday operations".

4. **A Staff Function -** Human resource management is a responsibility of all line managers and a function of staff managers in an organisation. Human resource managers do not manufacture or sell goods but they do contribute to the success and growth of an organisation by advising the operations department on personnel matters.
5. **Pervasive Function -** Human resource management is the central sub-function of an organisation and it permeates all the other management functions, *viz.* production management, marketing management and financial management. Each and every manager is involved in managing human resource of their departments. It is a responsibility of all line managers and a function of staff managers in an organisation.
6. **Challenging Function -** Managing human resource is a challenging job due to the dynamic nature-of people. Human resource management aims at securing unreserved co-operation from all employees in order to attain pre-determined goals.
7. **Development Oriented -** The individual goals of employees consist of job satisfaction, job-security, high salary, attractive fringe benefits, challenging work, pride, status, recognition, opportunity for development etc. Human resource management is concerned with developing the potential of employees, so that they derive maximum satisfaction from their work and give their best efforts to the organisation.

Human resource management (HRM) is the strategic and coherent approach to the management of an organisation's most valued assets - the employees, who contribute towards the achievement of business objectives individually as well as in teams. The terms "human resource management" and "human resources" (HR) have largely replaced the term "personnel management" as a description of the processes involved in managing people in organisations. Human Resource management is evolving rapidly. Human resource management is both an academic theory and a business practice that addresses the techniques of managing workforce.

## 1.4 Importance

The HR department is fast gaining recognition of being an integral part in the strategic management process. The importance of HRM can well be understood by the growing demand for HR professionals in the industry. HR plays an important role in creating jobs, providing unending supply of manpower, using manpower talent at the right place and creating conclusive environment for employees from all quarters of the society. Thus, it can be concluded that overall, HRM is considered important from the following viewpoints:

1. Organizational
2. Personal

3. Societal
4. National

## 1. Organizational

Human resources are the medium through which an organization fulfills its objectives. There management is an important function for effective performance. For an organization HRM is essential to keep a check on the costs, to get uninterrupted manpower supply etc.

1. **Talent management:** HRM is responsible for keeping an up-to-date skills inventory by managing the recruitments and performance management of employees.
2. **Cost Optimization:** HRM also helps the management in reducing costs by designing appropriate compensation policies that links rewards to performance. The trainings and development helps in enhancing employees capabilities thereby leading to raising the ROI of employees.
3. **Work culture:** By creating right attitudes and increasing team work and co-operation the organizational climate and culture become proactive and that nurtures overall development and welfare.
4. **Employer Branding:** Sound HRM policies help an organization creates a good brand image. The current global scenario believes that satisfaction in the internal environment is directly linked to organizational success.

## 2. Personal

HRM is also important from the perspective of the employees as it:

- ☆ Provides opportunities for growth.
- ☆ Enhances work life balances.
- ☆ Helps in identifying the latent potential of individuals through potential appraisals.
- ☆ Increases self-confidence.

## 3. Societal

By adopting labial practices, HRM has considerably helped in removing social barriers and constraints. It plays an important role in removing biases related to gender like different policies for male/female staff. It is also important in providing equal opportunities based on merits and generating a competitive spirit. By using job evolution techniques the labour issues are also managed quite effectively. Thus HRM also helps in reducing labor–management conflicts and creating a healthy environment.

## 4. National

Above all HRM plays an important role in the development of the country. It helps in:-

- ✰ Increasing employment opportunities.
- ✰ Development of skills and utilizing them.
- ✰ Raising the standard of living by providing performance oriented pay structures and providing a global organizational culture.
- ✰ Removes social and cultural differences.

The efficiency and performance of staff and their commitment to the objectives of the organisation are fostered by good human relationships at work. This demands that proper attention be given to human resource management and harmonious employment relations. The manager needs to understand the importance of good managerial practices and ways to make the best use of people. The promotion of good human relations is an integral part of the process of management and improved organisational performance.

## 1.6 Role of Human Resource Management

The human resources management team suggests to the management team how to strategically manage people as business resources. This includes managing recruiting and hiring employees, coordinating employee benefits and suggesting employee training and development strategies. In this way, HR professionals are consultants, not workers in an isolated business function; they advise managers on many issues related to employees and how they help the organization achieve its goals.

### 1.6.1 Working Together

At all levels of the organization, managers and HR professionals work together to develop employees' skills. For example, HR professionals advise managers and supervisors how to assign employees to different roles in the organization, thereby helping the organization adapt successfully to its environment. In a flexible organization, employees are shifted around to different business functions based on business priorities and employee preferences.

### 1.6.2 Commitment Building

HR professionals also suggest strategies for increasing employee commitment to the organization. This begins with using the recruiting process or matching employees with the right positions according to their qualifications. Once hired, employees must be committed to their jobs and feel challenged throughout the year by their manager.

### 1.6.3 Building Capacity

An HRM team helps a business develop a competitive advantage, which involves building the capacity of the company so it can offer a unique set of goods or services to its customers. To build the an effective human resources, private companies compete with each other in a "war for talent." It's not just about hiring talent; this game is about keeping people and helping them grow and stay committed over the long term.

### 1.6.4 Addressing Issues

Human resource management requires strategic planning to address not only the changing needs of an employer but also a constantly shifting competitive job market. Employee benefit packages must be continually assessed for costs to the employer. Tweaking the packages also provides an opportunity to increase employee retention through the addition of vacation days, flexible working arrangements or retirement plan enhancements. For example, in recent years many human resource professionals have oversaw the addition of preventative health components to traditional health plans for both employment recruitment and retention efforts.

## 1.7 Functions of HRM

From the definitions discussed above you can understand the basic nature of work related to HRM. The below figure depicts three-fold functions of HRM.

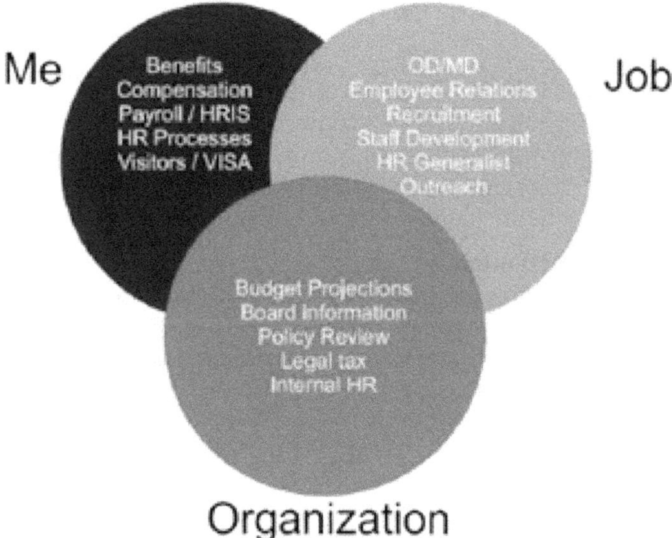


**Fig. 1.2: Functions of HRM**

Based on the above discussion HRM can be classified in the following categories:

(i) Functions for the Organization

(ii) Functions for Employees

### (i) Functions for the Organization

HRM is considered as both a Line and a staff function. This is because it involves the dual responsibility of designing as well as implementing the HR policies and procedures. The top management considers HR department as their part as it helps in the designing and management of employee policies keeping business goals as prime objective. These functions are also called managerial functions and include:

- Human resource planning that includes activities like projecting manpower requirements, bidet allocation for acquisition, development and motivation of employees. HR needs to get approval of management on these key decisions.
- Organizing HR includes activities like assigning jobs, establishing departments, delegating authorities, establishing communication systems etc. This function helps in streamlining the various jobs and tasks of employees with organizational vision and mission.
- Controlling refers to setting processes managing employee's performances, developing systems for appraisals, feedback and monitoring the organizational activities.

### (ii) Functions for Employees

Employee centric functions these are also called operative functions as these include the day to day activities related to acquisition, development and motivation of employees.

- **Procurement function:** refers to acquiring the right candidate for the right job in the organization. It is a continuous function involving recruitment, selection, succession planning and exit of an employee. Procurement function is important as it deals with filling the vacant positions by selecting the right candidate having required set of technical and soft skills.
- **Development Function:** As the term signifies, development function refers to enhancing the employee's skills, knowledge, behavior and overall capabilities to perform the assigned job effectively. Development involves activities training, management development, career planning, mentoring, coaching etc. Modern concept like OD, Learning organizations is focused towards overall development, employees as well as the organization. Development function helps in creating a talent pool that can be utilized for future business
- **Compensation:** It involves assigning wages and incentives to employees based on employee's performance. This function encompasses various activities related to reward management, performance incentives, etc
- **Motivation and maintenance:** The prime belief is that individuals are Lazy and therefore their performances behaviors need to be guarded. Thus performance management, motivation, and employee engagement activities are essential for achieving positive results. Maintenance is also an important function as it directly relates to employees well being. The major activities include employee engagement, grievance redressal systems, talent management, social security measures etc.

## 1.7.1 Business Function

Dave Ulrich lists the functions of HR as: aligning HR and business strategy, re-engineering organization processes, listening and responding to employees, and managing transformation and change.

At the macro-level, HR is in charge of overseeing organizational leadership and culture. HR also ensures compliance with employment and labor laws, which differ by geography, and often oversees health, safety, and security. In circumstances where employees desire and are legally authorized to hold a collective bargaining agreement, HR will typically also serve as the company's primary liaison with the employee's representatives (usually a labor union). Consequently, HR, usually through representatives, engages in lobbying efforts with governmental agencies to further its priorities.

Human Resource Management has four basic functions: staffing, training and development, motivation and maintenance. Staffing is the recruitment and selection of potential employees, done through interviewing, applications, networking, etc. Training and development is the next step in a continuous process of training and developing competent and adapted employees. Here, motivation is seen as key to keeping employees highly productive. This function can include employee benefits, performance appraisals and rewards. The last function of maintenance involves keeping the employees' commitment and loyalty to the organization. Some businesses globalize and form more diverse teams. HR departments have the role of making sure that these teams can function and that people can communicate across cultures and across borders.

The discipline may also engage in mobility management, especially for expatriates; and it is frequently involved in the merger and acquisition process. HR is generally viewed as a support function to the business, helping to minimize costs and reduce risk.

In startup companies, trained professionals may perform HR duties. In larger companies, an entire functional group is typically dedicated to the discipline, with staff specializing in various HR tasks and functional leadership engaging in strategic decision-making across the business. To train practitioners for the profession, institutions of higher education, professional associations, and companies have established programs of study dedicated explicitly to the duties of the function.

## 1.8 Personnel Management v/s Human Resource Management

Personnel management is a traditional form of employee administration. Since its primary subject is also employees, it was used interchangeably with HRM for a long time. Personnel management had evolved from labour issues, conflicts in countries like Britain and Australia, However it was regarded as linked with business strategies in US right from the very beginning. According to Goss, HRM has three principal distinguishing features s compared to PM. These are:

- ☆ It works beyond the set rules and contracts.
- ☆ Focus on strategy
- ☆ Individualization of employee relation

In general, personnel management is more of a control technique that take care of up keeping of system related to employees management. With time, employees get identified as resources, and were termed as Human assets. Incidentally HRM evolved to take care of issues concerning employee's motivation and development. Thus we can say that HRM has evolved out of personnel management but is more inclined toward employee's well being and attaining organizational goals.

Following are points of difference between personnel management and HRM:

- ☆ Personnel management is administrative that is it deals with pre-defined rules and procedures. Relation between employees and management is based on employment contracts. On the other hand HRM is concerned with building dynamic organizational culture. It deals with open contracts linked directly to business goals. HRM is employee centric and works on developing a positive work environment equipped with latest technologies, policies to give employees a performance oriented environment.
- ☆ Personnel management is a reactive function. It follows the set rules and guidelines and manages their day to day adherence by employees. HRM is a forward looking proactive approach that focuses on implementing innovative human oriented concepts. For example, Industrial relation, labour management functions are envisioned as key partners in organizational development. HRM practices, workers participation, whereas personnel management restricts itself to collective bargaining processes only.
- ☆ Personnel management is an independent management function is not linked to the other functions and processes unlike HRM. Personnel management is not involved in strategic management process. Its activities restrict up to payroll, compliances and employment laws. But HRM plays a significant role in strategic management and actively participates in accomplishing organizational mission and vision.
- ☆ Personnel management follows a collective approach; it works on standard practices that encompass all employees irrespective of their capabilities, competence and experience. On the contrary HRM work on individual development and management. HRM believed that every employee has different set of needs and therefore it focuses on individual development.

## 1.9 Human Resource Organisation

The organizational structure of Human Resources is usually very flexible and it reflects the immediate needs of the organisation. There is no modern organization, where the HR Structure has remained same for more than eighteen months.

The HR organizational Structure has to follow the needs of the organization, but it also has to allow the employees to operate smoothly and to deliver consistent results over the longer period of time. The HR organizational Structure is always tailored to the needs of the organization, but it is also necessary to find the models, which suit the organizational needs. The modern organizations use similar HR organizational structure to some extent.

The HR organizational Structure must fit with the imperatives defined by the need to keep costs low, provide operational excellence and to help in further development of the potential of employees. The HR managers must focus on enabling team co-operation and optimal processes and communication flows among team members. The HR organizational structure must be such that it ensures:

- ✰ Smooth co-operation among the employees.
- ✰ Smooth communication among employees.
- ✰ Clearly defined roles and responsibilities of each team member.
- ✰ Clear and easy-to-understand processes.
- ✰ Understandable structure for the managers in the organization.
- ✰ Potential for the future development of new initiatives and potential to improve the current HR processes.

As you can see, the conditions to be fulfilled by the HR organisational Structure are not an easy task. The HR organizational structures had undergone complex transformations so as to reach the current status.

The HRM functions evolved as a complete organization having the following unit's *viz.*, sales force unit, unit responsible for processing the known requests and units for product management and product development. This organizational structure seems very easy to implement, but to solve all the issues on the way of the implementation needs a lot of the power and effort.

### 1.9.1 The Resulting HR Organizational Structure

The current structure of the HR organization has the following three main functional areas:

(i) HR Front Office

(ii) HR Back Office

(iii) HR Centers of Excellence

#### (i) HR Front Office

Function is responsible for being a single point of the contact for the rest of the organisation. This concept is very hard to implement as the employees in HR Front Office must be able to find their position in HR department and they have to be recognized by the managers as a helpful source of information. The HR Front Office employees must be able to recognise the real needs of the internal clients and they have to be able to translate them into the requests for the HR Back Office and HR Centers of Excellence.

#### (ii) HR Back Office

Is responsible for providing efficient and errorless services to the external and internal clients. HR back office also makes sure that the activities of HRM are fully compliant with the legal environment.

#### (iii) HR Centers of Excellence

Are responsible for keeping processes, policies, products and initiatives updated, developed and fully competitive with the international practices.

## 1.10 Human Resource Management in India

The Indian economy has become an integral part of the global economy. Globalization has changed the way business was done in India. One of the important transformations has been of the personnel management in to human resource management. The transformation developed positive changes and started considering employees as partners of organizational growth. The key features of Indian companies have been:

- Family oriented business
- Less opportunity for innovation.
- Formal relationship between Management and employees

The Indian organizations started considering human aspects at work and traditional system of control were replaced by empowerment and participation. Still there are some basic challenges to be addressed by the Indian industries. These are:

- Developing a dynamic business organization.
- Ensuring employees engagement and long term commitment
- Motivation of employees.
- Enhancing employee's capabilities for raising productivity.
- Challenging the status-quo.

Thus Indian organization has to shift their focus from production oriented to quality driven organization. The current scenario of Indian organization has shown positive results. There has been an increase in white collar jobs, thanks to the IT boom. The workforce composition has also changed drastically comprising of baby boomers and generation Z. Organizations have become flatter and management has also shown liberal clues. As per John Rockfeller, the philosophy of HR professionals should follow "If you want to succeed, you should strike out to new paths rather than travel the worn out paths of accepted success".

## 1.10 Human Resource Development

Human Resource Development or HRD is defined as "organised learning activities arranged within an organisation in order to improve performance and/or personal growth for the purpose of improving the job, the individual and/or the organisation.

Human Resources Development is the composition that facilitates individual development, prospectively fulfilling the organisational or country goals. It's obvious that the personal development adds value to the person, and ultimately benefits the organisation and the country as a whole. In the context of business, the Human Resources Development consider workforce as an asset, whose worth is improved by development.

Human Resource Development (HRD) provides the basic foundation for helping the employees in developing their personal and organisational skills and knowledge. Human Resource Development includes such opportunities as:

- Employee training

- Career development
- Performance management and development
- Coaching
- Mentoring
- Succession planning
- Key employee identification
- Organization development

Human Resource Development lays emphasis on developing a superior workforce so that the organisation and individual employees can work on achieving their work goals. There are many opportunities in an organisation for human resources or employee development, both within and outside of the workplace. Human Resource Development can have both formal and informal aspects. It can be formal such as in classroom training, a college course or an organisational planned effort or it can be informal as in employee coaching by a manager. Human Resource Development covers all these activities to excel as a healthy organisation.

HRD includes the areas of training and development, career development and organisation development. This is related to Human Resource Management - a field that includes HR research and information systems, employee-employer relations, employee assistance, compensation /benefits, selection and staffing, performance management systems, and HR planning and organisation/job design.

## Questions

### Short Answers Questions:

1. Explain the concept and importance of HRM.
2. What are the functions of HRM?
3. Explain the organisational structure of human resource.

### Long Answers Questions:

1. Differentiate between personnel management and human resource management.
2. Explain the human resource management situation in the Indian industry.

# Glossary

**Talent Management -** It is a strategic HR function that involves developing special policies to Attract, motivate and retain employee of the organization.

**Procurement -** It refers to getting new employee having required skills in the employees to fill the vacant positions.

**ROI -** This refers to Return on Investment. It is a term used to signify the return from an employee in terms of productivity and value added to the organization for the cost incurred on him in form of salary, benefits, trainings and other expenses to develop his skills.

**White Collar Jobs -** It is a terminology used in corporate to refer to employees working at executive and above levels usually involved in management of organisation functions.

**Flat Organization -** These are companies that have less number of hierarchies and communication is direct.

**Industrial Relations -** Refers to the relation between employer, workers and government for the employment purpose.

**Payroll -** It is term used to denote the formal system of compensation and rewards to permanent employees of the company.

# Unit 2

# Introduction to Human Resource Planning

## 2.1 Introduction

Human resource planning may be defined as the process of assessing the organisation's human resource needs in light of organisational goals and making plans to ensure that a competent, stable work force is employed.

The efficient utilization of organizational resources human, capital and technological just does not happen without the continual estimation of future requirements and the development of systematic strategies designed towards goal accomplishment. Organisational goals have meaning only when people with the appropriate talent, skill and desire are available to execute the tasks needed to realise goals.

Assigning the right man for right job and developing him into an effective team member is an important function of every manager. It is because HR is an important corporate asset and performance of organizations' depends upon the way it is put in use. HRP is a deliberate strategy for acquisition, improvement and preservation of enterprise's human resources. It is a managerial function aimed at coordinating the requirements, for and availability of different types of employees. HRP is a forward-looking function and an organizational tool to identify skill and competency gaps and subsequently develop plans for development of deficient skills and competencies in human resources to remain competitive.

## 2.2 Human Resource Planning

Human resource planning is an ongoing process of systematic activities to achieve optimum use of an organisation's most valuable asset - its human resources. In other words, it is the process of systematically reviewing human resource requirements to ensure that the required numbers of employees with the required skills are available as and when they are needed. Human resource planning includes four factors:-

- Quantity - how many employees do we need?
- Quality - which skills, knowledge and abilities do we need?
- Space - where do we need the employees?
- Time - when do we need the employees? How long do we need them?

**Fig. 2.1: Human Resource planning**

Basically, human resource planning is the processes by which management ensures that it has the right personnel, who are capable of completing those tasks that help the organisation, reach its objectives. It involves the forecasting of human resources needs and the projected matching of individuals with expected vacancies.

A human resource (HR) department carries on a number of different functions, all of which are related to a company's employees. This can include recruiting talent, hiring workers, finding candidates for promotions and keeping tabs on future potential hires. The department's role in securing employees for a company is called human resource planning.

According to R. Wayne Mondy, human resource planning is the systematic process of matching the internal and external supply of candidates with job openings that a company anticipates over a certain period of time. Put simply, human resource planning is keeping an up-to-date compilation of candidates for company's future needs.

Human resource planning constitutes an integral part of corporate plan and serves the organisational purposes in many ways. For example, it helps organisations to)

(a) Capitalise on the strengths of their human resources

(b) Determine recruitment level

(c) Anticipate redundancies

(d) Determine optimum training levels

(e) Serve as a basis for management development programs

(f) Assess cost of manpower for new projects

(g) Assist productivity bargaining

(h) Assess future requirements

(i) Study the cost of overheads and value of service functions

(j) Decide whether certain activities need to be subcontracted In order to cope with such business needs:

- ✰ Suitable demand outline
- ✰ Excellent supervising and remedial measures
- ✰ Complete information about existing workforce and the external employment market.
- ✰ A perceptive about the resourcing functions in the business.

Applying a Human Resources Planning program would require sound preparatory work and comprehensive personnel records, which give accurate and objective data on all employees. We cannot make judgments on the supply of particular skills unless we have sufficient data on the skills possessed by existing employees. Building up full records requires both an effective system and determination to ensure the data is complete, up-to-date, and accurate. In addition, the information must be in a form that facilitates easy access during a review.

Plans should be set out with schedules of associated work force requirements, giving precise categories, skills, and levels for every function. These details will be necessary as a starting point when the questions of supply planning are tackled.

The decisions in an organisation can be made without considering HRP, but these decisions will devoid of the advantage of perceptive their inference. More over quantity of workers to be recruited will be laid down in unawareness of labour demand or management progression issues will be ignored.

## 2.3 Objectives of HR Planning

According to Sikula, "the ultimate purpose/objective of human resource planning is to relate future human resources to future enterprise need so as to maximize the future return on investment in human resources".

Based on the above definitions, the following objectives of HRP can be enumerated:

- ✰ Predict the effect of technology and future manpower requirement.
- ✰ Use the current HR in efficient way. If there is deficiency, plan for new recruitment and incase there is a surplus of manpower strategies like retrenchment and layoff can be adopted.

- ☆ Make certain that manpower is adequately supplied whenever required.
- ☆ Ascertain the different types of skills and competencies requirement depending on organizational strategies and future planning.
- ☆ Minimize excessive expenditure on manpower by evaluating the demand and supply ration.

When companies are global, an important challenge in garnering success is to respect other cultures and workforce environments and start forming a global profile or social consciousness. There are a number of reasons for the incorporation of human resource planning in an organisation. These are as under:

(a) It helps in discovering talented and competent workers and developing them to move up the corporate ladder.

(b) It ensures greater production by putting the right man in the right job.

(c) It helps to avoid a sudden disruption of an enterprises' production run by indicating shortages of personnel, if any, in advance.

(d) It helps to prevent under-utilisation of personnel because of over-manning and the resultant high labour cost and low profit margins.

(e) It provides information to management for the internal succession of managerial personnel in the event of an unanticipated occurrence.

## 2.4 Importance of HR Planning

- ☆ Upper management has a better view of the HR dimensions of business decision.
- ☆ Personnel costs may be less because the management can anticipates imbalances before they become unmanageable and expensive.
- ☆ More time is provided to locate talent.
- ☆ Better opportunities exist to include women and minority groups in future growth plans.
- ☆ Better planning of assignments to develop managers can be done.
- ☆ Major and successful demands on local labour markets can be made.
- ☆ Surplus or deficiency of employee strength is due to absence of planning.

## 2.5 Benefits of HR Planning

Most of the criticism about the HR planning arises due to its inclination towards the quantity features rather than qualitative attributes. So it's clear that the planning should not be focused on how many should be the work force but should be concentrated on the skills and capability. Moreover needed skills should be identified and proper plans should be prepared and included in the HR planning, to acquire it. But there are people who still believe that the HR planning should be based on the quantities. Human resource planning is considered as an integral part of business planning. Even though neither organizations nor employees can look into the future, making predictions can be quite helpful, even if they are not always accurate.

The basic goal of human resource planning, then, is to predict the future and based on these predictions, implement programs to avoid anticipated problems. Very briefly, humans resource planning is the process of examining an organisation's or individual's future human resource needs compared to future human resource capabilities (such as the types of skills among employees to be recruited or you already have). It also caters to developing human resource policies and practices to address potential problems for example, implementing training program to avoid skill deficiencies.

If the size of the employees recruited in more than which is required there will be a severe loss for the management. Even after not being used, the staff will have to be paid. However, if a miscalculation is made, resulting in the shortage of workers the company will be in all sorts of problems as it will struggle to meet deadlines for both sales and production.

Hence, the issues that come into the mind of an HR manager during the formulation of an HR plan can be pointed out as follows.

- What are the methods to be adopted to improve the efficiency by introducing a unique combination production, labor administration, and technological progress?
- How can be this correlated to the number of staff to be employed?
- What methods can be adopted to ascertain labour requirements?
- Is there any scope of adding more flexibility in work schedules?
- What are the methods that can be adopted for proper acquisition of needed workers?

These standards can be useful to any effort to identify labor necessities, for a new business venture, relocation, or rejuvenation of a plant or workplace.

## 2.6 Process of Human Resource Planning

The process of human resource planning involves several activities. It basically starts from the time Business plans are being drafted. These business plans provides the information regarding the number of employees needed for different jobs at different locations. This involves:

- Setting Objectives for HRP.
- Identifying key parties like the Top management, Line managers, HR departments.
- Establishing HR strategic mission and vision.
- Analyzing competitor's workforce and their capabilities.
- Implementing HR strategies.
- Evaluating results of planning process.

### 2.6.1 Phases

The HRP process consists of the following phases:

**Phase- 1 Identifying Manpower Demand**

HRP process starts by studying the organizational plans and business strategies for a given period usually between 3-12 months. This provides details of manpower requirement for achieving set goals. There are many factors that help in determining manpower demand like changes in government norms (apprenticeship acts, labor laws), technological changes (machinery, equipment's, international standards, quality) competition etc. For example, a company venturing into international operation will require expatriates and employees with international experience. This requirement will help in forecasting skills required in employees. . It has been observed that demand assessment for operative personnel is not a problem but projections regarding supervisory and managerial levels are difficult. Two kinds of forecasting techniques are commonly used to determine the organization's projected demand for human resources. These are Judgmental Forecasts and Statistical Projections.

(a) **Judgmental Forecasts -** Judgmental forecasts are also known as the conventional method. The forecasts are based on the judgment of those managers and executives who have intensive and extensive knowledge of human resource requirement. Judgmental forecasts could be of three types: Scenario Writing, La Prospective and Cross-Impact Analysis. Delphi technique Management jury.

(b) **Statistical projections -** Quantitative forecasting is often referred to as objective analysis while qualitative forecasting is called managerial or judgmental analysis. Quantitative forecasting can be characterized by one of the two basic techniques:

- ☆ Time series
- ☆ Relational-where the future is dependent on the direction of a variety of factors
- ☆ Work-study method
- ☆ Ratio- Trend analysis

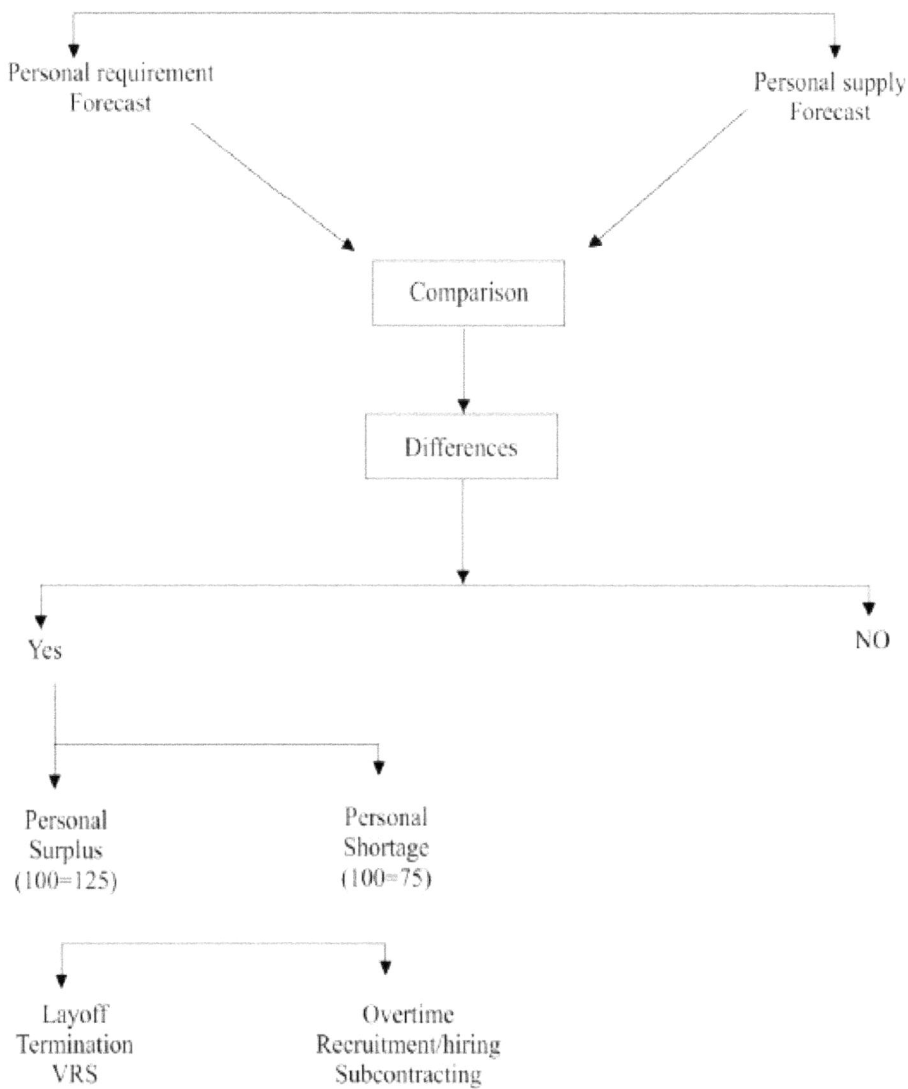


**Fig 2.2 Human Resource Planning**

### Phase- 2 Forecast Manpower Supply

After assessing the manpower needs of the organization, the next step is to forecast the supply of manpower from various sources. Here, the most feasible sources for finding the required manpower are assessed. It helps in ascertaining the quantity and quantity of people that can be sourced from different mediums. Forecasting of human resource begins with assessing the current human resource inventory of the organization. This activity is known as human resource audit.

After assessing the internal sources like data bank, promotions and transfers, etc, making allowance for absenteeism, internal movements and wastage and change in hour's external sources like placement agencies, labor contractors, etc are evaluated for getting the right type of employees. The supply analysis covers:

Skills Inventories - Information about non-managers. They are:

- Personal Data-Age, sex, marital status
- Special Qualifications-membership in professional bodies, special achievements.
- Company Data-Benefit plan data, retirement information, seniority
- Capacity of an individual-psychological test score, health
- Management Inventories: These include:
- Work history, strengths, weaknesses, promotion potential, career goals, personal data, number and types of employees supervised, total budget managed, previous management duties etc.

### Phase- 3 Comparing Demand and Supply of Manpower

This phase involves comparing the results of the previous two steps. The number of manpower required by the organization is matched with the number of manpower available through various sources. This brings out the potential gaps or surplus in manpower in an organization that become the basis for further action plan. Human resource planning therefore helps in striking a balance between the demand and the supply of employees which helps in reducing costs.

### Phase- 4 Action Plan

This stage deals with developing actions plans to fill the personnel gaps or to manage the redundant workforce. As the figure below depicts there are usually two sections of action plans one each for Filling the gaps and managing excess employees.

In case the employee is in surplus the management prime concern is cost. The surplus manpower leads to extra cost. Thus strategies like retrenchment, layoff, etc are decided.

In case manpower is short, and then suitable plans are made to fill these gaps through recruitments, trainings and development etc.

### Phase- 5 Review and Control

The final step in the HR planning is reviewing the action plans based on the results of demand and supply forecasting. This help in finding any discrepancies and driving the actions in the right direction.

## 2.7 Job Analysis

A job analysis is the process used to collect information about the duties, responsibilities, necessary skills, outcomes and work environment of a particular job. You need as much data as possible to put together a job description, which is the frequent outcome of the job analysis. Additional outcomes include recruiting plans, position postings and advertisements and performance development planning within your performance management system.

The job analysis may include these activities:

- Reviewing the job responsibilities of current employees.
- Doing Internet research and viewing sample job descriptions online or offline highlighting similar jobs.
- Analysing the work duties, tasks and responsibilities that need to be accomplished by the employee filling the position.
- Researching and sharing with other companies that have similar jobs.
- Articulation of the most important outcomes or contributions needed from the position.
- Its objectives include:
- Determination of the most efficient methods of doing a job.
- Enhancement of the employee's job satisfaction.
- Improvement in training methods.
- Development of performance measurement systems.
- Matching of job specifications with the person specifications in employee selection.

Comprehensive job analysis begins with the study of the organisation itself - its purpose, design and structure, inputs and outputs, internal and external environment and resource constraints. It is the first step in a thorough understanding of the job and forms the basis of job description which leads to job specification.

Job analysis is a detailed and systematic study of jobs to know the nature and characteristics of the people to be employed on the jobs. It involves collection of necessary facts regarding job and their analysis. Job analysis is a process by which jobs duties and responsibilities are defined and the information of various factors relating to jobs are collected and compiled to determine the work-conditions, nature of work, qualities of persons to be employed on job, position of job opportunities available and authorities and privileges to be given on the job etc. The main purpose of this analysis is to describe and define distinctions among various jobs in the organisation and their relationship.

The information collected under job analysis is:

- Nature of jobs required in a concern
- Nature/ size of organisational structure
- Type of people required to fit that structure
- The relationship of the job with other jobs in the concern
- Kind of qualifications and academic background required for jobs
- Provision of physical condition to support the activities of the concern, e.g. separate cabins for managers, special cabins for the supervisors, healthy condition for workers, adequate store room for storekeeper.

## 2.7.1 Job Analysis Methods

Job analysis is an important activity as it provides with detailed descriptions of the various jobs in an organization. The output of the activity helps in preparing Job description and job specification that are used managing, evaluating g and guiding employees towards organizational goals.

The three main methods of Job analysis are:

1. Observation method
2. Interview method
3. Questionnaire methods

### 1. Observation Methods

This refers to studying the jobs by watching the employees perform. The areas that are studies are:

- ☆ Behaviors
- ☆ Skills
- ☆ Leadership
- ☆ Knowledge
- ☆ Team working etc.

Job observation can be conducted by any of the following ways:

- ☆ **By Direct Observation:** This is a simple observation by a person of the day to day activities of an employees that includes his usage of machinery, his output or accomplishments, work environment, relationship with boss , communication patterns etc.
- ☆ **By Work Analysis:** This includes the time and motion studies *i.e.*, watching how much time an employee takes to accomplish a particular task. Another technique under this category is the micro analysis in which study is conducted through a camera in which every little aspect of an employee is recorded. This analysis helps in getting details regarding the challenges, constraints, job requirement, skills etc related with that job.

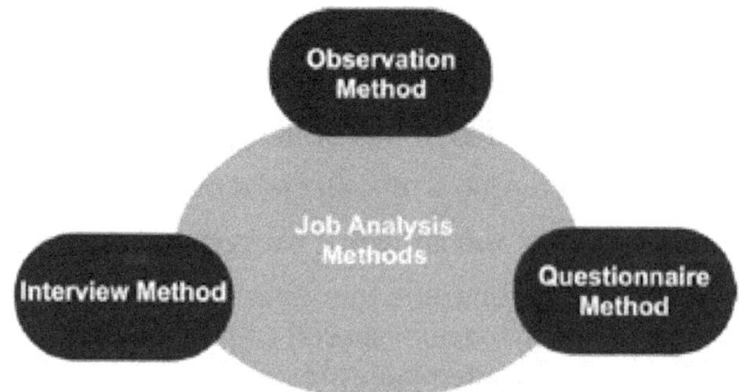


**Fig. 2.3: Methods of job analysis**

### 2. Interview Technique

Job analysis is also done by interviewing the employees on already set questions. The interview method is descriptive activity that involves discussion with the employee, his immediate bosses, and everyone who comes in contact with him for work purposes.

An important aspect related to this method is the role of the expert who conducts the interview. He has the responsibility of holding the discussion in one direction and eliminating any distractions and avoiding any arguments. Interview technique is extremely complex as the expert also has to deal with the employee's perceptions, apprehensions and queries regarding the need of the interview. Many time employees hesitate in giving details as they feel it's a performance appraisal method. Thus avoid providing concrete details.

### 3. Questionnaire Method

This method involves filling up the questionnaires by the employees. Questionnaires consist of various statements related with job. The question may be structures or u8bnstructures. To make it simple companies use structured and closes ended questionnaires. However, descriptive questionnaire with open ended statements are helpful in gathering every bit of details.

Using questionnaire is preferred over other methods as it is less time consuming and data can also be collected and analyzed easily. Generally a combination of questionnaires and interviews is also considered a good combination.

### 4. Functional Job Analysis Method (FJA)

This method uses a structured job analysis "schedule" and precise terminology in order to gather job related data. It is a descriptive method and is highly useful for recruitments, career planning and human resource planning.

## 2.8 Job Description

According to The British Institute of Management -"a job description is not intended to catalogue all duties involved with the result that an employee would feel justified in declining to perform any work not included in the description. It should be regarded as an outline of the minimum requirements of the job, thus preserving flexibility of operations".

Description of a job is an important written document that is prepared after in depth job analysis. It is a useful HR tool that guides the employee, his manager as well as other persons like customers, clients etc about the roles and responsibilities. It is a document that helps in segregating various jobs in an organization.

The importance and purpose of a job description is as follows:

- ✰ Enables career advancement in the company.
- ✰ Avoiding disputes among employees related to their duties and responsibilities.
- ✰ Determines of amount of payment to be made to employees.

- ✰ Increase of results by specifying the responsibilities.
- ✰ Grading and classification of jobs.
- ✰ Counseling, coaching and mentoring employees.
- ✰ Ascertain the promotions and transfers decisions.

**POSITION DESCRIPTION**

| | | | |
|---|---|---|---|
| **TITLE** | Administrative Assistant | **DATE** | January 29, 2015 |
| **COMPANY** | Optimus Performance | **DEPARTMENT** | Administration |

**REPORTS TO** President Stephen Goldberg

**COMMUNICATION WITH** President Associates, Office Manager clients suppliers

**SUMMARY OF OVERALL ROLE AND RESPONSIBILITY**

The purpose of this role is to support the smooth functioning of the company and assist the president by taking responsibility for all administrative task. This includes providing customer support, fulfilling client and associate requests and preparing marketing materials as required. By completing the following task well, this person will free the president to focus on sales activities and on delivery training and coaching services to clients.

**Fig. 2.4: Job Position Descriptions**

A job description specifies the following details about the job:

1. **Title of the Job :** Title of the job refers to the specifying identity of the job. For example Accounts officer, Sales Manager, Draftsman etc. The title of the job describes the nature of the job along with its level in the organization hierarchy.
2. **Location of Job :** Many companies have multiple offices and manufacturing units at different location like country, state and city. Thus job location is important as it communicates to the employee his location of posting.
3. **Job Summary :** It's a brief statement of 2-5 lines stating the general nature of the job. Job summary gives an overview that help in understanding type of job. It is also called job overview or nature of the job.

4. **Reporting Structure :** This section specifies the reporting details. It contains reporting to the seniors and the reporting from others. This section decides the seniority of the position in the group and the company.
5. **Duties & Responsibilities :** This section specifies in detail the work that has to performed by the candidate. It specifies the tasks, jobs to be fulfilled. This is the main section of JDs as it conveys the significance of the job. For an employer this section helps in identifying the right candidate during interview. It is also useful in managing performance.
6. **Working Conditions :** Every position in requires some specific work environment. For examples an accounts person, can work only if he is given a proper table, chair, computer, etc. Similarly, a salesman will be unable to perform if he is not given TA/DA or conveyance, laptop, telephone or mobile. Thus working conditions specifies the environment, machinery etc that is required to fulfill the given roles and responsibilities. It also mentions any sort of dangers attached to the job specially in cases of production related jobs.

Important consideration while preparing the job description:

1. Consider the business processes of the organisation.
2. Develop a process map.
3. Assign to the processes to the concerned employees.
4. Analyze the requirements in terms of machinery and equipments.

## 2.9 Job Specification

A Job Specification is a detailed illustration of the attributes, skill sets and prerequisites that should be possessed by a prospective employee. It outlines all the characteristics, which make the ideal candidate for a job. With the advent of online job portals and increasing job specializations, appropriate job descriptions are becoming ever more important for acquiring and reaching out to the right candidates. The job specification is an output of job description. It states the minimum acceptable qualifications that the newcomer must possess to perform the job satisfactorily. Job specifications serve as an important tool in the selection as well as evaluation process. The human resource department in consultation with different line managers develops these documents. The various elements of job specification are:

### 2.9.1 Significance of Job Specification

Companies can benefit from a well-written and informative job requirement design in various ways. The right job specification enables the recruiting organisation to reach out to the right candidates. It helps in acquiring relevant applications for the job and reduces the time for short listing for interviews. It is a means of proper marketing and advertising of the job opening of the company in various mediums like print, internet etc. Job specifications aid the human resource in placing the applicants in the appropriate role and position for a job.

**Contents of Job Specification:**

### 1. Education

This states the minimum formal education necessary to perform the job. It specifies the level of education that is acceptable fir the job, foe example for a helper; it is usually high school or Intermediate. For a manager at least a graduate. Education also specifies the need for professional qualifications like ITI, MBA, Hotel management etc.

### 2. Training

Many jobs require some specific type of expertise. This section mentions the training needs like ERP, CRM and Six Sigma etc.

### 3. Experience

This section mentions the minimum experience required for performing the job. For example" required MBA fresher with 0-6 months of experience". It is important as it communicates the level of job and the expectations of management from the candidate.

### 4. Skills

Job specification provides detail about the type of skills required in the candidate. It can be related communication, leadership, manual or mental skill. For example for a typist "should have a speed of 60-75 WPM". For a analyst should be well versed on MS Excel etc.

### 5. Physical Attributes

Personality factors such as personal appearance like age, weight, height, eyesight etc.

### 6. Psychological Characteristics

For high-level jobs, the ability to assume responsibility is an essential prerequisite. Also includes emotional stability, maturity, initiative, drive and sociability.

## 2.10 Retaining Your Highly Skilled Staff

Human resource planning is the process by which an organisation ensures that it has the right number and kind of people, at the right place, at the right time, capable of effectively and efficiently completing tasks that will help the organisation achieve its overall objective. Human resource planning provides the organisation with a detailed plan of action designed to achieve the organisation's objectives and comes up with the information about the number of workers needed to meet those objectives. Without a clear cut planning, estimation of human resource need remains nothing but a guessing game on papers.

Though issues about retention were not in the forefront any discussions in recently, every business were keen to learn about the progression of resignation and assist itself in retaining its highly skilled staff. Thus, every business should:

- Observe the degree of resignation
- Determine the motive for it
- Find out the cost that the business will have to forfeit due to resignations.

- Evaluate this cost with other resembling businesses
- Devoid of this perceptive, organization may be oblivious of the eminence of workers lost.

This expenditure will be rendered by the organisation through the invoice for separation, recruitment and induction and also by a loss of enduring competence. By understanding the character and degree of resignation, methods can be adopted to correct the condition. Comparatively inexpensive and easy remedies can be sought out after finding out the reasons for the exit of employees.

## 2.11 Managing an Effective Downsizing Programme

Downsizing of labor force is a frequent matter of concern for managers. Safeguarding of enduring benefit of the business organisation and lessening the strength of the workforce without affecting the business is a responsive matter. It happens to be more complicated along with the pressure of production and sales targets on the management attached with employee apprehensions.

HRP facilitates the business by taking into account:

- The kind of labour force predicted till the conclusion of the exertion.
- The merits and demerits of the diverse routes which enables success of the precise long-term plans of the business.
- The effectiveness of retraining, posting and reassignment.
- Pointing out the suitable recruitment levels that will be required.

The result of such an assessment can be put forward to the senior managers, so that the expenditure advantage of different means of reduction can be considered and the period taken to congregate the objectives can be recognized.

## 2.12 Strategic Human Resource Planning

Strategic Human resource planning is a significant part of strategic human resource management. It connects Human resource management straight away to the strategic plan of the business. Most middle to big sized businesses generally has a strategic plan that directs it in the fruitful combination of its tasks. Organisations regularly fulfill the financial plans to make sure that they accomplish organizational objectives and but the labor force plans are not given predominance.

Even a tiny business with as little as 10 employees can build up a strategic plan to make judgments about the opportunities. Depending upon the strategic plan, a business can put up a strategic HR plan that will enable it to create HR management choices to maintain its bright future path. Strategic HR planning is also significant for the financial planning such as yearly budgets as the business can predetermine operational expenditures of staffing, training etc.

A complete human resource strategy has a crucial part to play in the success of an organization's general strategic plans, clearly demonstrate that the human resources function completely appreciates, and sustain the course in which the business is advancing. It also maintains extra precise strategic aims undertaken by different departments such as marketing, financial, operational etc.

Essentially, Strategic HR Planning give importance to intelligent use of the workforce, if a business looking forward to attain standard to long term objectives. This can be achieved by:

- ☆ Retaining the precise people in specific positions
- ☆ Observance of the exact combination of skills and aptitudes
- ☆ Encouraging the employees with factual approaches and performances
- ☆ Attitudes and behaviour is developed in the right way

Strategic HR planning predicts the future HR management needs of the organisation, after analyzing the organisation's current human resources and the future HR environment of the organisation. The analysis of HR management issues external to the organisation and developing scenarios about the future are what make strategic planning a distinguished activity.

The basic questions to be answered for strategic planning are:

- Where are we going?
- How will we develop HR strategies to get there successfully, given the circumstances?
- What skill sets do we need?
- The overall purpose of strategic HR planning is to:
  - ☆ Ensure adequate human resources to meet the strategic goals and operational plans of your organisation - the right people with the right skills at the right time
  - ☆ Keep up with social, economic, legislative and technological trends that impact on human resources in your area and in the sector
  - ☆ Remain flexible so that your organisation can manage change if the future is different than anticipated

The HRP strategy should demonstrate that effectual planning of workforce matters will make the business easier accomplish its wider intentional and organized goals.

Strategic human resource planning consists of a number of objectives, but all the objectives must be closely aligned to overall business goals in order to be effective. Human resources executives must demonstrate the employer's return on investment through strategic activities. Strategic activities are forward-thinking processes that support business growth through recognition of the value of human capital.

It's the prime objective of the HR manager to make sure that the HR Strategies in the business are incorporated with the wider organizational goals. Moreover, it should be taken care that it the strategies implemented are acceptable to others also. In order to attain these objectives, HR mangers should

- ☆ Communicate with all the stakeholders and detail them about the proposed strategy.
- ☆ Promote and build up supporters of the strategy through effective communication.

- ☆ While communicating with others concentrate on the benefits that are going to be achieved by implementing the strategy and illustrate examples for the same.
- ☆ Assessment of the extent of commitment to the strategy across the business will increase the success rate.
- ☆ Customary response of the plans should be given with the help of employee newsletters, presentations etc.
- ☆ Wherever applicable, put together the strategy proven results, as that is easy to be monitored and evaluated.
- ☆ Implement the strategy in the orientation process for the work force - in particular for the senior level.

## 2.13 Strategic Human Resource Planning Model

The strategic HR planning has to undergo the following steps:

1. Evaluate the present HR competence
2. Predicting the future HR needs
3. Gap assessment
4. Initiating HR strategies to prop up the business objectives.

### 1. Evaluation of the Present HR Competence

In accordance with the approved strategic plan of the business, the evaluation of the present HR competence of the business is the first and foremost step of the HR planning. The awareness, proficiency and aptitude of the present work force are to be determined .The method involves the preparation of an index according to the skills of each worker. The willingness of the employee to accept more duties and responsibilities according to his personal career development plans should be acknowledged and motivated.

### 2. Predicting the Future HR Needs

The second step is to predict the future HR requirements according to the strategic objectives of the business. Practical prediction of human resource needs involve the assessment of the labour demand and supply. To deal with the related issues an HR manager should find out the answer following:

- The number of workers needed to meet the strategic objectives of the business?
- The specific employment vacancies that are to be filled in the recent future?
- The aptitudes that the workers should possess?
- Influence of the external environment on the HR requirements of the business?

## 3. Gap analysis

The subsequent step is to decide the gap between vision of the business about its future and the present condition. The gap analysis comprises making out the quantity of workforce, expertise and aptitude obligatory in the future in contrast to the existing condition. HR management practices must be observed strictly to recognize conducts that can be enhanced or the innovative practices that are essential to sustain the organisation's capability to progress.

To deal with the related issues an HR manager should find out the answer for the following:

(a) What are the innovative jobs that should be created?

(b) What are the innovative skills that will be compulsory?

(c) Does the present work force have these innovative skills?

(d) Are the personnel's in various positions posted according to their strengths?

(e) Does the business have the adequate managers?

(f) Are the existing HR management practices sufficient for future requirements?

## 4. Creating HR Strategies to Maintain Organisational Strategies

The HR strategies which help the gathering of the future business needs are

### (a) Restructuring Strategies

These strategies include the:

- ☆ Dropping the strength of the work force by termination or attrition.
- ☆ Reorganizing the jobs to generate perfectly designed jobs.
- ☆ Reorganizing job divisions to be more competent.

Termination of employees provides instant outcomes. Usually, there will be expenses related to this approach according to the service contracts. Attrition is another method to lessen staff strength. The feasibility of this alternative relies on how much instantly the reduction of staff is essential. The jobs executed in the business are essentially restructured so that work of the leaving employee is also done.

### (b) Training and Development Strategies

These strategies include:

- ☆ Giving proper training for the workers to do the consigned new roles.
- ☆ Giving the existing employees the opportunity for development to set up them for upcoming jobs in the business.

Training and development wants can be conferred in different ways. One method is by the business paying the employees to improve their expertise. This may entail the sending the worker to receive training or certificates or it may be able by means of on-the-job training.

### (c) Recruitment Strategies

These strategies include:

- ✰ Recruiting fresh personnel's with the talent and aptitude that the business shall require in the future.
- ✰ Bearing in mind all the accessible options for purpose fully supporting employment openings and encouraging appropriate aspirants to submit an application.

For strategic HR planning, every recruitment process the organisation should be looking at the needs from a strategic viewpoint. The recruitment strategy must be to discover somebody who will be capable to align with the modifications that the business will plan for the future.

### (d) Outsourcing Strategies

This strategy includes usage of employees outside the organization to do the work:

Several organisations rely upon the external staffs by engaging them for certain skilled jobs. This is above all supportive for carrying out definite, expert responsibilities which does not require a permanent work. So a number of organisations subcontract HR conducts, assignment work or accounting. For instance, payroll could be prepared by an outside organisation rather than a organization staff, a temporary assignment may be completed by a professional or particular expertise. For example a legal guidance may be acquired from an external legal firm.

### (e) Alliance Strategies

These strategies include:

- ✰ Functioning jointly to persuade the kinds of courses presented by educational establishments.
- ✰ Functioning with other organisations to organize future managers by giving out in the growth of talented personnel's.
- ✰ Allocation of the costs of training for the team of workers.
- ✰ Permitting workers to trip to other organizations to increase proficiency and skill.
- ✰ Ultimately, the strategic HR planning process could guide to indirect strategies that go ahead of the business.

## 2.14 Planning the Total Workforce

Planning the entire workforce is an exceptionally significant action, which finally aim to implement the HR functions of succession planning, training and development effortlessly since they turnabout the existing labor force. There are normal transformations in the existing work group as fresh applicants adhere and older ones depart.

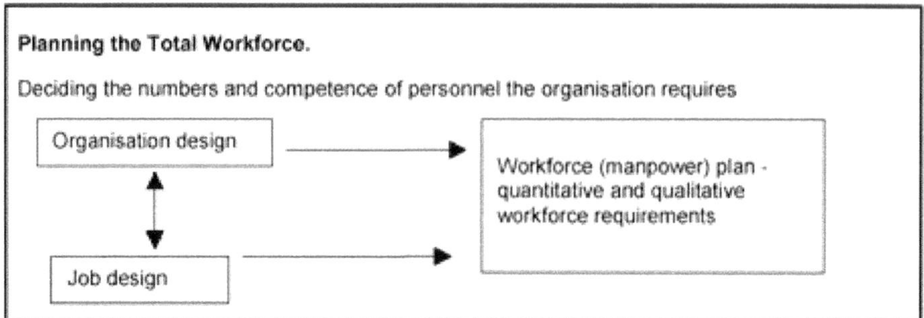


**Fig. 2.5: Planning of total workforce**

Shaping upcoming business necessities, mainly those concerning to work force necessities, symbolize one of the most demanding responsibilities human resource managers have to face.

The sketching out of a workforce plan is an important element of every human resource strategy and one of the anticipated products of human resource manager's actions. In spite of this, manpower or labor force planning, along with succession planning, has only newly had the benefit of revival in reputation. The failure of several businesses to build up and execute labor force planning is somewhat problem-solving of the need of strategic planning itself.

Strategic goals of each business are most important for its success and labor force planning works according to it. In order to convene the company's strategic goals the procedure should be ingenious and should identify the employee's individual and expert features. This preparation is also accomplished to build up the strategies of fulfilling the business needs. It is a systematic progression that gives the manager a representation for taking human resource decisions in accordance to the business functions, strategic plan, allocated funds and a set of preferred labor force proficiencies. So the workforce planning can be stated as an organized process that is incorporated, arranged, enduring and reliable in character.

According the requirement of the businessman workforce plan shall vary from easy to intricate. It makes out the necessity of the business in terms of prospective workers to meet organizational objectives. This brings out the determination of quantity and skills needed along with the time schedule they will be required. Workforce planning might be for a division or subdivision or for the entire business to offer productive out comes to the business; the success of these plans rely upon its assimilation with much broader management strategies. The labour force planning also initializes the improvement of strategies to fulfill the business needs and it includes the taking appropriate actions to magnetize and maintain workers of mandatory skills along with upholding most favorable figures according to the organization's needs.

Besides the labor force planning, it has to be guaranteed that organizational structure along with the jobs guarantee the proficient rendering of services and administration of the organization as a whole.

Suggested proceedings:

- Shaping of proper organisational structure to maintain the strategic goals.
- Arrangement of the employment according to the key activities.
- Preparing a workforce plan proposed to maintain the organisations strategic goals.
- Bring together workforce profiles, discovering the designated groups, an record of existing workers competencies, competencies needed in the future and acknowledged gaps in competencies.

## 2.15 Investing in Human Resource Development and Performance

Conventional concepts of career planning, performance appraisals, compensation management and workman development should be restructured according to the idea, features and assignment conclusions as in the HR plans, strategies and procedures.

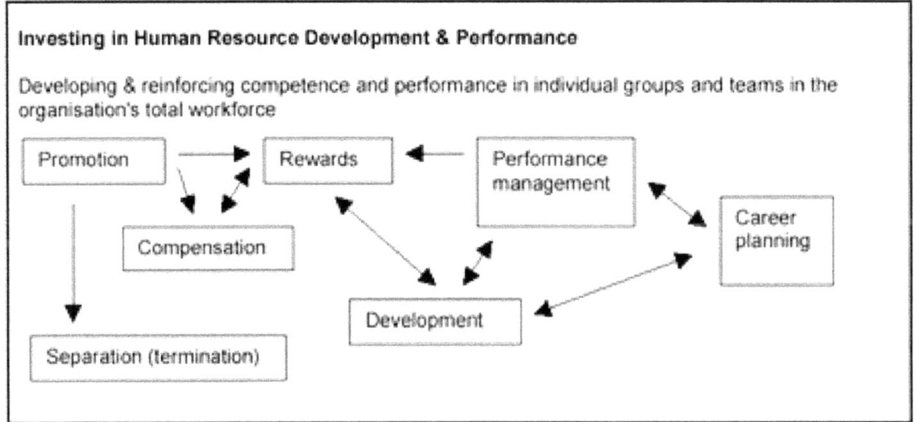


**Fig. 2.6: Investing in HRD and performance**

Development reactions will intend to amplify business skills. The appliances of business skills and the behavioral essentials are all add to an organization for successful performance. New perception such as ultimate learning and identifying previous learning should shape an important constituent of the way of investing in employees. Strategies for diverse organization differ according to its development pace. Workers retention strategy also follows identical procedure.

At the time of workforce planning process if there are indications of limited encouragement or career progress in an organization and the development features are also limited, organizational strategies proposing the retaining of the workers will be advisable. But in organizations which have significant growth and expansion will have an opposite strategy. Every significant proposal is taken for the person, group and organization for ensuring maximum performance levels and business development.

It is vital that at a personal level, mainly for superior staff, that they experience that their growth desires are approved and they be encouraged by providing the skills to do their works. At a group level, it characterizes the personal capability to coordinate his work with others and support personal and group skills to facilitate the business goals.

Compensation strategies have an objective to support the functions of the organization with the method of giving compensation to its people, giving the obligatory incentives and enthusiasm to the workers. Its elements are the blend of the basic pay, bonus, revenue sharing, share preferences, and a variety of suitable benefits, generally according to the market or competitor rules and the organization's capacity to compensate.

**Suggested actions**

Sufficient investment in HR guarantees a sturdy base for the business and contented employees. This can be achievable by accurate exposure of suitable measures, policies, and procedures according to employment development and learning, promotion and employment assignments, compensation management, retirement etc.

## Questions

### Short Answer Questions

1. Explain the need for human resource planning.
2. What is the importance of HR planning?
3. Differentiate between Job specification and job description.

### Long Answer Questions

1. Explain job description. State the main points that any job description document mainly stresses upon. Also explain its significance.
2. What is job analysis? Explain various methods of job analysis.
3. Explain the analyzing factors of manpower requirements.
4. Explain the process of developing employment plans.

# Glossary

**ERP -** It refers to Enterprise resource planning. It is software that streamlines the activities of all the departments of an organization. It's an important skills required in process oriented organizations.

**Job Description -** It is a detailed list of roles, duties, responsibilities, reporting relationships, working conditions and supervisory responsibilities related with a particular job. It is an important description used in performance appraisal, training and recruitment activities.

**Absenteeism -** It refers to the absence of employees from work on frequent basis. Absenteeism is regarded as a negative sign in an employee's behavior.

**TA/DA -** These stands for Transport allowance and daily allowance. These are reimbursements made by company to specific employees who have to move out for official purpose.

# Unit 3

# Human Resources and Business Growth

## 3.1 Introduction

The small-scale business sector has been a major contributor to our country's growth. Since the time of independence, this sector has huge growth prospect in India with its wide range of products. With 40 percent share in total industrial output and 35 percent share in exports, the small-scale industrial sector in India has a bigger role to play in increasing the country's Gross Domestic Product (GDP).

Many small businesses operate with no employees. One person handles the whole business with perhaps occasional help from family or friends. Hiring someone to help is a big decision as one has to worry about payroll and benefits. Employers are also worried about a host of problems that can arise from personality conflicts and loss-of-control of all the processes in running the business.

A survey of personnel functions of smaller firms found that functions like finance, accounting, production, and marketing all take priority over personnel management. Personnel management is considered as a second most important management activity next to general management/organizational work. Most small business owners manage the personnel functions themselves as the work force is small. They do not develop sophisticated personnel systems. But as a small business grows, personnel practices begin to emerge in importance and more systems and process are being developed.

## 3.2 Human Resource in Small Business

A very basic definition of small-scale industrial classified it into two categories-first one used power with less than 50 employees and second was the other group, which did not use power and had an employee strength of more than 50 but less than 100. However, over the time, other criterions were introduced like capital resources invested on plant and machinery buildings. They became the primary criteria to differentiate the small-scale industries from the large and medium scale industries. An industrial unit can be categorized as a small- scale unit if it fulfils the capital investment limit fixed by the Government of India for the small-scale sector.

The latest definition, which is effective since December 21, 1999, describes small-scale industries as the functional unit of a business which satisfies the following conditions:

There should be a limit of Rs. 10 million on the investment in fixed assets like plants and equipment's either held on ownership terms on lease or on hire purchase basis. These units cannot be owned or controlled by any other industrial unit.

Substantial work has been undertaken in the field of human resource management (HRM), as it applies to large organisations. However, for small businesses same thought-process cannot be frequently applied. A small business lacks adequate systems to ensure efficient management of human resources. Further, most small businesses are the product of their owners, whose personality and personal involvement dominate. From the experiences of various small business owners who have experienced growth in their business, it was found that motivating and retaining good staff was a critical bottleneck in their business growth. The point of awareness for many came from managerial training programs. However, this was tempered by their beliefs and the growth cycles of their companies. Prior to change being possible these owner-managers needed to develop skills and competencies in leadership, coaching and management before effective delegation and team building could take place. These findings are linked to the existing body of knowledge relating to HRM.

Many small businesses operate only with an owner. They cannot justify a dedicated HR resource. As a result, the owners manage their business obligations internally. These obligations can include anything from writing a new contract to tackling health and safety issues. To maintain a competitive business, it is vital that the owners not only understand these obligations but also know how to manage them efficiently.

The business environment today is more competitive than ever, so having a dedicated workforce is an invaluable asset to any business. Getting the right team in place starts with the recruitment process – it has to be ensured that the management really understands the role that has to be filled and the type person both the boss and the team can work with. By getting this right, the team gels so well that not only they work hard for themselves, they work for each other too.

Once the selections are made, it is good to spend time getting to know the employees and letting them to know the boss too. It has to be found out what makes them tick and learn how to work with them to get the best results. The employees should have the leadership that they require with a set of achievable goal backed up with great motivation and training; thus giving them all the tools they need to make their own contribution to the success of the business.

The boss sets the strategy for the business and works with the teams to deliver on this strategy whilst constantly keeping an eye on the bottom line. Complying with employment obligations on top of all that can sometimes just feel like an added job to do on the list.

It is not surprising when it comes to managing obligations - whether it is calculating an employee's maternity leave or providing them with a work contract - some bosses believe they need to draw in external help. They find employment obligations to be complicated and time consuming. But it can be made much simpler by taking control of the obligations personally. That would not only save time and money but also ensure much more control and confidence in running the business day-to-day.

The most critical challenges faced by today's small business owners in the management of their human resources fall within the following three categories:

1. Retention of qualified staff.
2. Prioritizing on becoming an administrative organisation instead of being a craft or a promotional organisation.
3. Taking proper care of the special needs of a diverse workforce.

Following are the factors that increase the complexity of this challenge:

- First, a number of small businesses are experiencing a limited number of available local applicants.
- Second, there is frequently a lack of accessibility due to limited or non-existent transportation options to reach the workplace (mostly situated in rural or far-flung urban areas).
- Third, small businesses are often limited in terms of the wage and benefit packages they can offer, particularly as they compete with larger businesses.
- Fourth, small businesses often lack advancement possibilities for existing staff. There are either no growth opportunities or the existing staff lacks the necessary technical expertise or administrative experience to fill the positions when they become available.
- Fifth, the transient nature of applicants with limited qualifications may make it difficult to retain them.

Human resource management function of a small business is not of the same size or complexity as those of a large firm. A business that has only two or three employees may face important personnel management issues. Indeed, the stakes are very high in the world of small business when it comes to employee recruitment and management. A business cannot run with an employee who is lazy, incompetent or dishonest in nature or dealings. A small business with a work force of half a dozen people will have a worse impact on its process by such an incompetent employee as compared to a company with a work force that numbers in the hundreds (or thousands). It is rightly said that, most small business employers have no formal training in how to make hiring decisions".

Though, SMEs (Small and Medium Enterprises) include organisations of different sizes with changing extent of intricacy in executive practices they habitually are considered as one unit. The small business owner should take into account many factors prior to the hiring of a fresh employee. Katzell argued its difficult generalise the case of one organisation and expect that it's applicable to all the other organisations. He recognized definite proportions, such as size of the firm and the degree of relations of organisational associates, which determine the difference and intricacy in organisational practice and conducts. Going after Katzell's proposal, organisational size along with the strategic inference of growing size should be taken into account while considering suitable stages of proper HRM practices appropriate to every organisation .Blau pointed out two wider tendencies as organisations enlarge in size. The first is growing allotment of labour, resulting to more horizontal and vertical separation. The second is –according to the increase in size, at first quickly and then slowly the differentiation enhances.

Talents of prospective employees should be matched with the company's needs of a small business. Efforts to manage this can be accomplished in a much more effective fashion if the small business owner devotes energy to defining the job and actively taking part in the recruitment process. However, human resource management task does not end with the creation of a detailed job description and the selection of a suitable employee. The hiring process signals the start of HRM functions for the small business owner.

**Fig. 3.1: Small business owner**

Small business consultants strongly urge that even the most modest of business enterprises should implement and document policies regarding human resource issues and functioning. In order to manage their enterprise's employees, some small business owners also need to consider training and other development needs. The need for such educational supplements can range dramatically. A bakery owner, for instance, may not need to devote much of his resources to employee training, but a firm that provides electrical wiring services to commercial clients may need to implement a system of continuing education for its workers in order to remain viable.

To establish and maintain a productive working atmosphere for the employees of the company, the small business owner needs to work on developing the same. Employees only become a productive asset to the company when they feel that they are treated in a fair and professional way. The small business owner, who communicates business expectations and company goals to his employees, provides adequate compensation offers meaningful opportunities for career advancement and anticipates work force training and developmental needs have better chances to prosper in his business. These practices help in providing meaningful feedback to his or her employees too, which leads him to be more successful than that owner who is neglectful in any of these areas.

The administration of human resources in SSI involves:

### 3.2.1 Staffing

The Human Resources Management (HRM) function includes a variety of activities and key among them are:

**Fig. 3.2: Staffing**

- ✰ Developing personnel sources
- ✰ Recruiting potential employees
- ✰ Equal Employment Opportunity
- ✰ Hiring Issues
- ✰ Interview Guidelines
- ✰ New employee orientation
- ✰ Termination

Thus, HR is involved in deciding what staffing needs are and whether to use independent contractors or hire employees to fill these needs. Recruiting and training the best employees, ensuring they are high performers, dealing with hiring and new employee orientation issues and ensuring that, personnel and management practices conform to various regulations are also important objectives to be taken care of.

### 3.2.2 Training and Development

Taking staff out of work for a training course can be difficult in a small business where people usually fulfill more than one role. But in most industries training is essential if you want to keep up with your competitors.

**Fig. 3.3: Training**

The best training for a small business is short sessions focused on a specific need. It usually keeps its focus upon the following:

- ✰ Initial Training.
- ✰ Training Needs Analysis.
- ✰ Preparation of training programs.
- ✰ Management and supervisory development.

### 3.2.3 Policies and Procedures

HR is involved in managing the approach to employee benefits and compensation, employee records and personnel policies. Usually small businesses have to carry out these activities by themselves because they cannot yet afford part-time or full-time help.

However, they should always ensure that their employees are aware of personnel policies, which conform to current regulations. These policies are often in the form of employee manuals, which all employees have.

Fig 3.4: Policies and Procedures

- Instituting health and safety standards
- Privacy Issues
- Handling of employee grievances
- Preparing employee handbooks
- Setting employee work hours
- Union relations

### 3.2.4 Record Keeping

For many small businesses, the biggest problem is, not knowing where to start with business records-so none are kept. An appropriate record-keeping system can determine the survival or failure of a new business.

Fig. 3.5: Record keeping

For small business, good record-keeping systems can increase the chances of staying in business and the opportunity to earn larger profits. Complete records will keep in touch with business's operations and obligations and help to see problems before they occur. This includes;

- Maintaining individual employee files
- Maintaining all of the department records
- Preparing departmental reports for top management

### 3.2.5 Wage and Salary Administration

Wage and salary administration management is about a range of interconnected processes that aim through both financial and non-financial means to address what employees value in the employment relationship. It includes:

Fig. 3.6: Salary

- Job analysis
- Preparation of job descriptions
- Pay surveys

- ☆ Compensation
- ☆ Employment Taxes
- ☆ Fringe benefits

## 3.3 Literature Review and Hypotheses Development

Formal practices of HRM process are important for the organisational functions. Here, the word formal refers to suggested or prescribed practices where the emphasis is on those practices, which are generally approved in the literature as appropriate for the various HRM areas examined. In this regard, it can be said that this hypothesis extends beyond documentation and standardisation of procedures, roles and instructions to include legitimate sources of recruitment and the use of specialists for training. From Katzell's contention, SMEs (small and medium enterprises) of varying sizes should demonstrate various levels of formality in their HRM practices.

Hornsby and Kuratko examined HRM practices of small firms keeping their studies consistent with Katzell's propositions. They examined three size categories and reported increased sophistication in practices with firm's growth. Roberts, Sawbridge and Bamber argued that the limits of informality become apparent in firms with 20 or more employees when informal networks of recruitment dry up and when informal styles of management communication are stretched. Jennings and Beaver noted that at this size the owner becomes overextended and needs to delegate responsibility to more professional management. Wilkinson studied relations of employees in the business and argued that employment relations in SMEs are characterised by informality. They insisted that formal control systems and communication strategies are almost non-existent in this set up. He maintained that emphasis on rules and procedures is out-dated in an environment where owners have to make speedy decisions in response to market pressures.

Several similarities were found among small and large firms in many areas of HRM practices by Golhar and Deshpande. These contrasting views make it difficult to understand existing HRM practices in SMEs and to prescribe appropriate practices for these firms. The result derived from these contrasting studies gives an approach, which is a "one-size-fits-all" approach to HRM training and advice for SMEs.

Following hypotheses are suggested; keeping in line with the argument that HRM, practices will become formal with firm's growth;

- ☆ **$H_1$:** As the firm grows a greater variety of formal recruitment sources are introduced.
- ☆ **$H_2$:** As the firm's size increases, multiple selection methods intensify the screening of candidates.
- ☆ **$H_3$:** The application of formal employment procedures at the managerial level lags behind those processes at the operational level for smaller firms.

## 3.4 Recruitment and Selection

With the changing times, the recruitment of employees has become both easier and at the same time more challenging too. The advent of the online world of job boards, recruitment websites and social media increases the likelihood of finding world-class candidates. The online world supported by our matching criterion also gives so many candidates to consider for a position that weeding through applications becomes time consuming and hectic activity.

Recruitment is the absolute method of making out of the potential applicants in this manner motivating and persuading them to submit an application for a specific job or jobs in the establishment. It is an encouraging act as it entails attracting individuals to apply. The idea of application calling procedure is to create an inventory of suitable candidates from whom, appropriate choice of the most apt person is done.

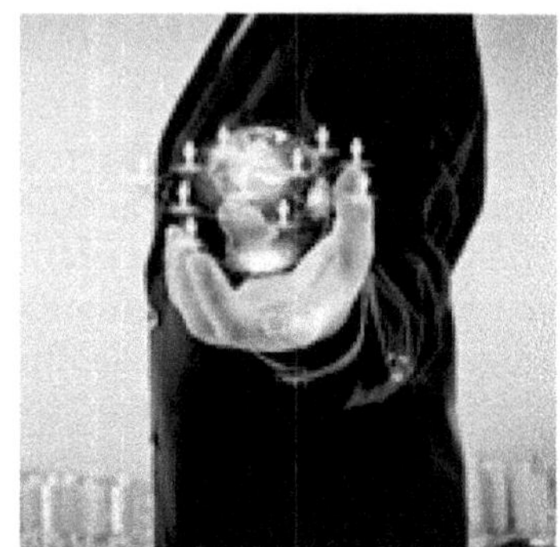

**Fig. 3.7: Online world**

After identifying the sources of human resources, finding out the potential candidates and persuading them to submit an application for the vacancies in the organisation, the management has to perform the function of selecting the right employees at the right time.

The selection process involves judging candidates on a variety of dimensions, ranging from the concrete, measurable like years of experience to the abstract, and personal like leadership potential. To do this organisations rely on one or more of a number of selection devices, including application forms, initial interviews, reference checks, tests, physical examinations and interviews.

**Fig. 3.8: Selection**

All selection activities, from the initial screening interview to the physical examination if required, exist for the purpose of making effective selection decisions. Each activity is a step in the process that forms a predictive exercise, which managerial decision makers seek to predict which job applicant will be successful if hired. Successful in this case means performing well on the criteria the organisation uses to evaluate personnel.

Finding the best possible people who can fit within the culture and contribute within the organisation is a challenge and an opportunity. Once we find, retaining the best people is easy if we do the right things right. It is important to have a good organisation structure, but it is even more important to fill the jobs with the right people. Staffing includes several sub-functions:

(a) Recruitment or getting applications for the jobs as they open up

(b) Selection of the best qualified from those who seek the jobs

(c) Transfers and promotion

(d) Training those who need further instruction to perform their work effectively or to qualify for promotions

Owner-manager of small firms prefer informal sources of recruitment and as the firms grow, the skills and abilities required for various functions and processes may not be available from the familiar and informal recruitment sources leading them to look for alternative sources. Thus, a greater variety of formal recruitment sources would be used to attract suitable candidates. Hornsby and Kuratko reported extensive use of newspaper advertisements, government and private employment agencies, employee referrals and unsolicited applications among small firm recruitment practices. Barber et al. also did the same research and found that his findings were consistent with the earlier studies where there was increased use of formal hiring procedures as firms grow.

As firms grow, multiple selection techniques were introduced along with taking interviews in order to reduce errors in selecting employees recruited from sources, which were unfamiliar to the owner-manager. Golhar and Deshpande demonstrated that one-to-one interviews are the most popular selection techniques in both small and large firms, with large firms also likely to use written tests and panel interviews in the selection process. Hornsby and Kuratko reported extensive use of application forms and reference checks in the selection process as firms increased in size. Barber et al. found from his studies that larger SMEs tend to rely on objective qualifications and they also use a greater number of different selection procedures in making hiring decisions. Rowden found that technical skills and positive work ethics received high priority in the selection processes of several of the successful small manufacturing and processing firms.

The adoption of formal employment procedures at the managerial level will lag behind that at the operational level for small firms, as owner-managers prefer to employ the few managers that are required from amongst family and friends. As these familiar sources become no longer able to cater to the specialist skills required this difference is expected to diminish with further growth of business.

## 3.5 Training and Performance Appraisal

According to Dunn and Stephens, Training refers to the organisation's efforts to improve an individual's ability to perform a job or organisational role whereas Development refers to the organisation's efforts (and the individual's own effort) to enhance an individual's abilities to advance in his organisation to perform additional job duties. Thus, training provides knowledge & skills and development provides efforts required to perform the job.

**Fig. 3.9: Business Training**

Organisations and individuals should develop and progress simultaneously for their survival and attainment of mutual goals. Employee training is a specialised function and is one of the fundamental operative functions of human resource management too. Training improves changes and moulds the employee's knowledge, skill, behaviour, aptitude and attitude towards the requirements of the job and the organisation. It also bridges the difference between job requirements and employee's present specifications.

Management development is a systematic process of growth and development by which managers develop their abilities to manage. It is a planned effort to improve current or future managerial performance. MacMahon and Murphy found out that SMEs rarely carry out formal training-needs analysis and have no systematic approach to training. Training often is perceived as an unaffordable luxury involving not only course fees but also the cost of unproductive labor. Owner-managers are of the view that that training results in highly specialized staff, whereas, multi-skilled workforce is required to cope with the highly flexible nature of jobs. In contrast, Hornsby and Kuratko reported the use of a variety of training methods in small firms with on-the-job training being the predominant method. Most of the owner-managers of small business firm perform almost all business activities by themselves or they honestly administer the performance of these actions. And hence, they encompass the responsibility of all the key business actions, like employee training, where they instruct their workers in the manner they want their work to be carried out.

People differ in their abilities and their aptitudes. There is always some difference between the quality and quantity of the same work on the same job being done by two different people. Performance appraisals of Employees are necessary to understand each employee's abilities, competencies and relative merit and worth for the organisation. Performance appraisal rates the employees in terms of their performance.

These appraisals are widely used in the society. The history of performance appraisal can be dated back to the 20th century and then to the second world war when the merit rating was used for the first time. An employer evaluating their employees is a very old concept. These are an indispensable part of performance measurement and they check the progress of employees towards the desired goals and aims.

The modern view of the organisations to compensate according to the work delivered, points out their concentration on individual performance leading to effective performance management. If the process of performance appraisals is formal and properly structured, it helps the employees to clearly understand their roles and responsibilities and give direction to the individual's performance. It helps to align the individual performances with organisational goals and also review their performance. Performance appraisal takes into account the past performance of the employees and focuses on the improvement of the future performance of the employees.

## 3.6 HRM in Business Growth

Throughout much of the 1990s and early 2000s, CEOs were rewarded for cost reduction end cost containment. To compete in the global arena, Boards demanded dramatic improvement in efficiencies. Companies such as General Electric, Motorola etc. fuelled the trend by reporting 10 times or higher returns on investments (ROI) from their Six Sigma programs. With a clear focus on the bottom line and tantalizing ROIs, companies invested heavily to get lean, outsource nonstrategic work, improve quality and introduce better products.

HR demonstrated in the earlier period that it could be a strategic player in efficiency initiatives by sourcing work and talent for better advantage, outsourcing routine work and moving some talent engagement costs from fixed to variable. A sensible but attentive concept of connecting the HR strategy to the development strategies of the business must be carried out together by the HR groups and their department equivalents, with the intimate concern of line leaders. Improvement in business plan has produced fresh demands to relocate HR efforts, roles and precedence in order to discover new basis of development.

In minor industrial firms, it turns into the liability of the CEO or owner to draw attention, maintain and inspire brilliant juvenile and precisely able individuals. In big multinational companies, the influence of individuals on the aggressiveness of businesses has become clear and HR professionals are made to help in creating strategic clarity across the organisation.

Over the past 20 years, human resources strategy has evolved through various forms. Twenty years ago respected companies like IBM, Microsoft and Tata pioneered early efforts to create HR strategies, which aimed to link HR functional strategies with the business priorities.

As companies seek new ways to define pragmatic but innovative people strategies, another transition is underway that will directly support business agendas, which work towards driving stock prices and expectations for growing shareholder value. A short list of items that receive management attention over the period of time is termed as an agenda. For an HR Agenda to be effective, business challenges demand focus on its perception. To close the gap between the strategic-HR haves' and have-nots', practitioners need a thoughtful but practical approach to achieve the following:

- There has to be a set of clear choices, to connect the people priorities to business priorities
- The line ownership for HR outcomes should have a detailed clarification and a contract specifying the responsibility that HR people will have is necessary.
- People should be guided, so that a proper allocation is done to the most important activities of the business.
- The energy of the workforce should be driven upstream towards the things that produce more impact.
- The organisational and systems framework should be defined for the HR function across all its business units.
- People should be encouraged and energised with a sense of purpose.

## 3.7 The Growth Gap

Growth can transform the market valuation of a company because it attracts talent, which creates the capital to grow faster. Yet many companies have a significant enterprise growth gap: the difference between the sum of the forecasted business units' growth goals and the overall enterprise target.

In other words, it is the difference between the actual delivery of core businesses and the expectations of the CEO and top management team about the business growth. This is generally described as a financial shortfall, but the root cause is a gap in the enterprise's capabilities and processes to identify and exploit new opportunities beyond the reach of the core businesses. For example, in a 10Billion multi-business chemical company, its core businesses were able to generate 5 percent growth though current expertise, but the board of directors and CEO desired 10 percent growth in order to increase stock price and market capitalization. To bridge the 500Million revenue gap, the prime responsibility of the CEO and Executive Team was to create new families of products and services by generating new businesses as per the size and reach of company.

Most companies believe that closing of revenue gap will be achieved by doing better at and more of the same things. However, good strategy, understanding of markets and technologies and analytical support will not ensure that the growth gap will be closed. The most significant pillars of success rest in resolving the following:

- Existence of an enterprise growth gap.
- Planning of the leadership team unified around growth goals and their leading path.
- Designing of an organisation so as to achieve the growth that has been identified.
- Keeping the source of growth clear.
- Finding out whether the innovation process can deliver new businesses that will achieve the growth goals.

When HRM is actively involved in driving the organisation to achieve a positive answer for these questions, it will be aligned closely to the CEO growth agenda too. This will require HR to adapt to the new vision by adding the development of leaders who can identify and execute new ideas in new spaces. This will also lead to the realignment of the organisation systems and structures for growth to its strategic competencies.

### 3.7.1 Closing Growth Gap

After studying high growth companies such as Procter & Gamble, UPS and Medtronic, Oyster International two major success factors in closing the growth gap were identified.

First, these companies identified New Growth Platforms (NGPs). From there they went on successfully creating New Growth Platforms (NGP), which were basically the material for new families of products and services typically outside the reach of the current businesses.

Second, the companies exhibited a common set of characteristics in business innovation, leadership and organisational design to sustain the flow of NGPs. These characteristics include:

- ☆ Efficient and credible chief growth officer who acts as the in-charge of the new unit.
- ☆ Believing and acting on the fact that the team is more important than the idea.
- ☆ NGP units are independent units but with a strong interdependency to the core businesses.
- ☆ Having a disciplined, systematic and repeatable process for screening, selecting and building NGPs.

This process guides and leads HR to the forefront. The HR advances to grab the opportunity to be a player in the company's growth agenda and not just sit on the sidelines. HR can help design and build a more agile organisation capable of identifying, designing and iterating a pipeline of NGPs. We can say that, HR can really help in creating a culture and organisation that is focused, innovative and highly disciplined.

## 3.8 Four Roles of HRM

HR Role in the organisation is quickly changing. HRM has been developing them for several decades but the function is not that mature, as many people would expect. The real concept of Human Resources responsible for the human capital inside the organisation evolved in the early 70s of the 20th century.

HR Role in the organisation is changing from the industry to the industry and it also depends a lot on the country, in which the organisation operates. The dream position of every HRM Department is to become an important deciding authority. It gets more and more responsibilities and participates on strategic initiatives in the organisation.

David Ulrich's HR Model is well known for introducing mainly the aspects of human resources with the highest value added. The main contribution of the David Ulrich's HR Model was the start of the movement from the functional HR orientation to the more partnership-oriented organisation in HRM Function. Business Partnering is not possible to implement without a major shift in the HR organisation. The benefit was a more responsible and flexible organisation of Human Resources, which allowed many HR Professionals to become real respected business partners.

Strategic Partner is about alignment of HR activities and initiatives with the global business strategy and it is the task of the HR. Sometimes, it sounds easy to implement Strategic Partnership, but it needs a lot of effort from Human Resources.

Change Agent is a very important area of the Ulrich's HR Model. Change agent is about supporting the change and transition of the business in the area of the human resources in the organisation. The role of Human Resources is to support for change activities in the change effort area and ensuring the capacity for the changes.

Administrative Expert changes over the period of time. In the beginning, it was just about ensuring the maximum possible quality of delivered services, but nowadays the stress is put on the possibility to provide quality service at the lowest possible costs to the organisation.

Employee advocate is a very important role of Human Resources. The employee advocate knows what employees need and HRM should know it. The employee advocate is able to take care of the interest of employees and protect them during the process of the change in the organisation.

All the HR Roles defined by Ulrich are essential for the success of the whole HRM Function. The world around us is changing and the HR Roles have to change as well. In the past, the HRM was responsible only for developing the processes, which assure the top quality delivered to the organisation. Nowadays, the HRM has to deliver even more. The HR roles and responsibilities have to take the high-level recognition of the organisation and they need to be adjusted to make a full fit.

The HR Roles have to be adjusted to:

- Keep HRM function focused on tracking and implementing new trends in the industry.
- Keep HRM function focused on helping the line management to implement improvements.
- Keep HRM function focused on operational excellence.
- Keep HRM function responsible for developing the human capital potential in the organisation.

## 3.9 Human Resource Planning in Business

Human resource planning is a key activity in business. It includes workforce planning, employee development plans, building basic employee benefits and salary programs, training and development, hiring and firing of employees etc.

### Steps

Even though it's a corporation of one right now and the growth may not necessarily include additional employees in the near future, there are the steps one will want to start while planning HR needs. These steps will also work for a growing business that already has employees. It has to be ensured that one spends enough time and effort to develop strong human resource planning that clearly identifies the present and future needs, policies, gaps, goals and actions.

**Preparation of a Forecast:** This is where HR anticipates how many employees it will need in the future. The longer it remains in business, the more accurate the number will be because it can look at the growth in the past and forecast based on those numbers. For example, if the business has grown by 25% in the past three years, there is a good chance that it'll need at least one employee in the very near future. However, if the experience in the industry suggests that this could be a quiet year, it may include that in its planning, too. So the business's history and industry experience can be a guide in Step One as the considerations are made for the supply and demand of the products.

**Develop HR Inventory:** This is fairly easy for most small businesses. However, if it's a growing business with a couple of employees already on board, then that should figure into the planning, too. As the business grows, this step becomes more important as the HR takes into account the employee turnover factor into the calculations, too.

**Develop a Job Analysis:** This could be the toughest part of the process; HR will be needed to figure out what each person will do even though they haven't been hired. It will give an idea about what the people on board will be getting to do and it can also help in training the current employees if their job description does not match the job description that they'll have when the business will have more employees. It should be kept in mind that any current employees will be managing the ones hired after them so they should be included in the job analysis.

**Prepare Comprehensive Plan:** For hiring the employees advertising the job and interviewing the applicants is the most important thing to do. There needs to be comprehensive plan for that. This is going to be the most time consuming of the steps, but it is the most critical. This step will give the path the HR Plan will take in order to integrate new employees into the business successfully. It should include budgeting for future wages and it should also include training techniques to bring current employees up to the upgraded skill level. These 4 steps will help in planning the small business HR to achieve success. A good human resource planning and management process results in creating good rapport and better communications levels between employees and the management. It also results in individual employee development plans and ensures that there is low employee turn-over, low absenteeism and low safety incidents. Well-designed and competitive employee compensation programs (including basic employee benefits), results in high morale, good quality output and efficient operations by well trained employees and also ensures satisfied customers.

## 3.10 Outsourcing HR Reaps Benefits for Small Business

A study on small business organisations show that their owners spend between seven and 25 percent of their time in handling employee-related activities which mostly comprise of paperwork. This results in a lot of unproductive use of their time. But by opting for outsourcing of some or all of their employee-related functions — such as payroll, benefits, healthcare or recruitment and retention — these business owners can look forward on focusing other important matters of business, which ultimately helps them to improve their productivity and even save some money.

**Fig. 3.10: Benefits for Small Business**

One would imagine that outsourcing is something that concerns only large companies. Well there is some truth in it, but small businesses too can reap the benefits of outsourcing by understanding that the benefits that accrue are substantial. This has been well appreciated by many small-scale companies. Estimates suggest that about 47% such companies are actively considering outsourcing administration, 44% are looking at outsourcing IT services and another 23% at distribution and logistics facilities. Rising costs and vendor pricing are the primary reason for even smaller companies to be favourably disposed towards outsourcing.

There are in fact a number of advantages that accrue to a small business when it decides to outsource. For one there is a marked decrease in your capital costs as your fixed costs get converted to variable costs. At the same time it enables more efficient use of scarce capital.

Outsourcing frees up time and resources, which can be better utilized in providing value to customer. Moreover, new projects can be taken up far more easily. That enables small firms to benefit from technology and cost advantage which would otherwise be unavailable to them. Not only that, the companies that one outsources to are far more adept and managing the risks involved in their areas expertise, a great relief for harried small firm owners.

However, there is a great deal of apprehension among small firms particularly those that outsource to countries like India about quality control. The way to deal with this is to check if all-important processes are in place. An important reason why outsourcing has become the new mantra for many small firms is that the spread of internet and broadband has led to a proliferation of small firms employing fewer than 100 people. The fact that they are strapped for time makes outsourcing almost a matter of no choice for them.

The prime mover in small companies having a serious look at outsourcing is the fact that technology has brought the price of outsourcing down and the sheer convenience and consequently economy that it offers have convinced the most tightly managed firms to wake up and act accordingly.

Outsourcing as a matter of fact is being intertwined in everybody's life. It helps them cut costs, get access to great coverage plans at competitive rates and thereby keep employees motivated, attract skilled and experienced professionals and bring about streamlined functioning of their business.

Outsourcing of HR services of small and medium-sized companies leads to long term benefits for the company. They can benefit enormously from the services of a service provider, as it helps these companies by securing competitive rates for some comprehensive and attractive coverage plans for the employees and the business itself. The expertise of the outsourcing agency in all aspects of HR management helps the client organisations to focus purely on their core responsibilities by keeping the burdens of outsourced HR functions away from their owners. The areas in which an outsourcing service provider can help with respect to employee benefits include:

- Life Insurance
- Initiating good work/life program

- Personal accident cover
- Risk coverage due to long term and short-term disability
- Designing a flexible spending account plan
- Credit Union
- Educational assistance for employee up gradation
- Appropriate retirement services

Reliable service provider review the current benefit plans to see if any improvement is necessary, thereby ensuring employee satisfaction.

## 3.11 Changing Organisational Structures/Work Patterns in "SHAMROCK ORGANISATIONS"

Another development is the Changing organisational structures and work patterns. The upcoming companies, according to Charles Handy, will be a shamrock organisation. The shamrock organisations will have the following three elements:

1. A few and important core group of experts, professionals and managers
2. A collection of sub-contractors who manufacture goods and services that the core group does not
3. An emergent group of provisional and part-time employees who are hired to give specialised services or to assist at maximum workloads.

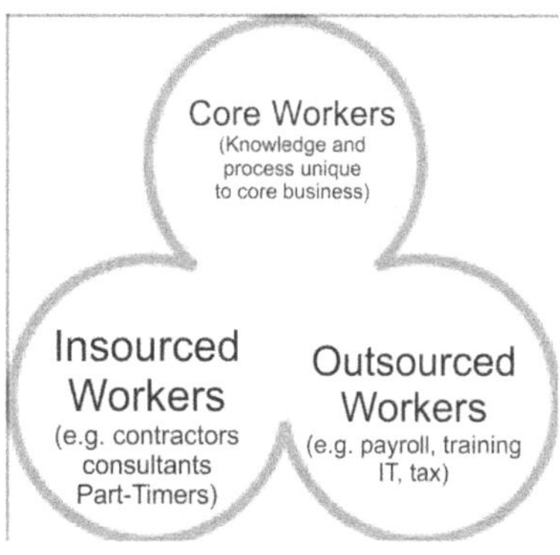


**Fig. 3.11: Shamrock Organisations**

Industries in many industrialised nations are more and more becoming shamrock organisations. Shamrock organisations will come out in budding countries in Asia and the Pacific. This is by now occurring in the NIEs. For example the Singapore National Trades Union Congress (NTUC) in a research on service structure and its effect on union membership (NTUC 1996) has proved that the shamrock organisation is by now forming in Singapore. The diminutive core of experts, professionals and managers will necessitate being the centre for human resource management. More ventures in the administration and teaching of part-time and casual labours will also be compulsory.

Technical change, particularly information technology and telecommunications and rivalry in the quick moving aggressive international market have transformed work organisations and work patterns. The manufacture of goods and services has turned into flexible and modified in spite of being bulk produced in extensive production lines. Unchanging mechanization relating recurring tasks is being substituted by flexible mechanization. On-line quality control is substituted by end-of-line checking. As an alternative of breakup of responsibilities, rising utilization is made of groups and multi-skilled employees. Managerial decisions are being decentralised to point of manufacturing and sale.

The organisational ladder is restructured with centre layers of management removed. The gap linking those in management of the organisational leadership and those accountable for production and delivery of goods or services is lessened. Due to these transformations in working models; the role of employees has widened with a resulting necessitate for a wider variety of skills. For example the cellular manufacturing operators (CMOs) in Motorola in the United States are likely to carry out their own jobs along with gathering the knowledge of variety of extra assembly functions. This is to take full advantage of the flexibility of the groups. Moreover, they ought to be largely self-overseeing. CMOs toil jointly in comfortable teams and are in charge for planning their job, controlling excellence and discovering innovative conducts to get better work procedures, diminish imperfection and cut down succession time.

On the other hand in the banking sector the job of customer service representative (CSR) has been extended further than a sequence of recurring errands of approving cheques, responding to usual questions and serving customers to resolve the account balance. CSRs currently also suggest a suitable blend of products and services of the bank such as mutual funds and allowances to clients. Workers at present require a lot of broader balance of skills than in the earlier case of the conventional slower moving hierarchical establishment with apparent chains of order with the "thinkers" at the apex and "doers" at the base.

The workforce has turn out to be more mobile. The major impact on work patterns is the upcoming relationship organisations or virtual organisations. A virtual organisation is a business, which with help of information technology connects different industries, vendors, consumers and sometimes the competitors and helps them to share technology and the expenditure and entrée to each other's markets. The virtual organisation has an exceptionally minute core with a lot of resources supported from outside and having no physical set up. Virtual offices are upcoming as companies and are promoting the cyberspace and electronic expertise to reduce expenses like rentals and to enhance efficiency. In these offices the employees don't attend the offices but keep in touch through the technology by using the devices which can send and receive messages, faxes and emails. Telecommuting is a type of the virtual office in which the employees do their job at home or any place other than the office. According to Richard Nolan, a Professor of Business Administration from Harvard University forecasts that there is a possibility that the virtual office will become conventional rather than atrial within three years' time.

The growth of virtual organisations has HRD inference. Virtual organisation needs employees that are extremely consistent and skilled, talented to appreciate and acknowledge the novel types of information, flexible and can work resourcefully with others. The workers should not only need the technical aptitudes but also the aptitude to learn how deal with uninterrupted and fundamental transformation of virtual businesses. New types of training that are flexible, according to the demand and interactive should be made for the workers of virtual organisations.

The changes in work pattern are occurring in the industrial nations. The developing countries of the Asia pacific region are also being influenced by these work pattern changes. Human resource development guidelines and program will have to be altered according to these changes.

The ASTD's 1990 report listed out the seven groups of skills that the employer desires.

## 1. Knowing How to Learn

This can be considered as the fundamental skill. By virtue of this skill the workers can rapidly learn more skills easily. This skill comprises the ability to gather, examine, arrange and implement the information. It includes the methods, approaches and information that ease dispensation of information. It is also the skill to use suitable knowledge as well as the ability to relate it in new circumstances at work. This talent consequently facilitate the workforce to get used to fresh demands at work.

Now, the learning has become an element of working life with aggressive work pressures and varying technology. Moreover, the accessibility, quantity and density of data have amplified. Employers perceive the aptitude of acquiring the knowledge as the solution to retrain and regular learning. Above all the skills helps the effective usage of the innovative know-how and work in accordance with the organisations strategic goals and competitive challenges.

## 2. Reading, Writing and Calculation

The conventional working styles often a routine process or continuous interaction with machines. So there are chances of Illiteracy and innumeracy in the work force as this cannot be determined easily. But in the modern workplace there is a need for high interaction with modern computerised machines that need a good reading skills and computation. Advanced numerical skills are needed with the implementation of different statistical approaches. We all know that writing is preliminary step for the communication with consumers, to document competitive dealings or effectively moving new ideas into the workplace.

It has been seen that most of the workers spent many hours during their working hours in reading different charts, graphs manuals etc. for handling different tasks. Then also the writing stays as the fundamental method of communicating policies, procedures and concepts. Moreover different tabulations are made every day in order check upon inventories, organize report on production levels, compute machine parts or provisions etc.

In adequate skills will lead to low productivity, high disaster rates and pricey production faults. All above the employer's capability to fulfil the strategic goals and the effort to become competitive will be weakened.

## 3. Communication Skills

Communication is the centre heart of the even functioning of any organisation. These skills are so important because they help to win the heart of customers and help to retain them. Plunging novelty, contributing to quality circles, settle on conflicts and giving significant advice all pivot on effectual communication skills.

The communication consumes a lot of hours of the workers and their success on the job is related to their communication skills. It's evident by the recent studies that only job knowledge comes above the communication skills as an aspect for workplace accomplishment.

Most of the business heads have a perspective that a deficit in these skills will result in a cost of millions for the employers every year in the form of low productivity and errors.

## 4. Adaptability Skills

Most of the business organisations are now a day's giving an extra importance for the employee who is problem solvers and an innovative thinker. Competitive advantage is commonly attached to a company's capability to innovate rapidly. This capability is according to the quantity of skills that employees have to free themselves from ordinary thoughts for a creative leap.

To be a successful problem solver one should possess aptitude to solve an individual problem along with aptitude to solve existing in a group problem. Above all he should possess the practical capability to relate these individual and group problems and solve them. Hence the Cognitive skills, group communication skills and problem handing out skills are critical to victorious problem solving.

Creative thinking is the skill to make use of variety of thoughts, ending up with something different and new in order to envisage, anticipate or shape new grouping of ideas to accomplish a requirement. Usually the creative thinking in the workplace is related to creative problem solving. This requires an effective team work and the assessment of the problems in an innovative way and the finding out of fresh solutions to active problems. The organisations capability to attain its strategic goals depends upon the capability of the work force in problem solving.

## 5. Developmental Skills

Personal management skills are base for the building up of a good self-esteem, a determined work life and even organisational efficiency. Now often the workers are increasingly called upon in order to take decisions regarding the production or sales or any function related to them which increases the positive sense of self-worth. This also helps the employees to be motivated. Individual employee's not having proper motivation or goal setting aptitudes will create recurring errors and quality issues or it can delay change.

## 6. Group Effectiveness

It's seen that while doing work the employee will have to interact with people and hence will require excellent interpersonal, teamwork and negotiation skills. The interpersonal skills is the ability to umpire and poise proper behaviour, manage with unwanted behaviour in others, soak up stress, deal with uncertainty, listen, encourage self-confidence in others, delegate the responsibility and cordial interaction. It's obvious that these aptitudes are needful for the effective conciliation of conflicts, which can be considered as the part of the work life. The negotiating capacity can be considered as the skill to detach the people from the crisis, to concentrate on benefits not positions, to work out compromise for a common achievement.

Interpersonal and arbitration skills are the foundation of a winning teamwork. Teams, which are more and more being used, are prearranged in the workplace so that suitable aptitudes and talents can be shared to achieve vital responsibilities and goals. A featured teamwork is an outcome when the team members distinguish and deal with various distinctive behaviours and when each has an idea of the customs and approaches that former group members symbolize. Above all the team members also should be aware of group dynamics that develop and transform as the team move towards its goals.

### 7. Influencing Skills

At its most basic level, leadership means that an individual can persuade others to act in a definite way. Often the employees are required to persuade his work group and to give an idea of what the entire establishment or the specific task necessitates.

## 3.12 Training in Small Industries

Training provision varies considerably in small businesses with size being a significant factor. The British small business sector has a long-standing reputation for poor training levels but West head and Storey found considerable variations between the smallest (micro-enterprises) and larger SMEs. Smaller firms conducted their training internally with a focus on informal skills learning. Larger SMEs tended to obtain more external, formal training with a goal of obtaining recognised qualifications. A bakery owner, for instance, may not need to devote much of his resources to employee training, but a firm that provides electrical wiring services to commercial clients may need to implement a system of continuing education for its workers in order to remain viable.

- ☆ Small industries need to consider training and other development needs in managing their enterprise's employees. The need for such educational supplements can range dramatically.
- ☆ In a survey of 6000 randomly selected SMEs found significant differences in owner/manager attitudes and approaches towards training needs of family and non-family employees in their businesses. The needs of family members were seen in terms of firm-specific HRD issues such as succession planning whereas training for non-family employees was focused on individual career needs.
- ☆ Owner/managers were mostly positive towards training but did not regard it as a critical element in overall business strategy. As studied about medium sized businesses it was observed that in over two-thirds of the sample, their owner/managers were heavily involved in HRD issues. But, despite the increasing levels of complexity and formality found in such organisations, there was still a big number which overlooked HRD issues. HR managers took charge of training and HRD decisions in only 26 per cent of cases. All respondents claimed that they used training plans according to the decided budgets and saw a strong link between their firms' specific training needs and sustainable competitive advantage. However, in spite of taking

interest in training of employees, most owner/managers did not accept the importance of this approach in their overall business strategy. Instead, they took an alternate approach perceiving the HRD needs of their workforces as an expensive affair and believed that the training of non-family members was an organisational expense which would affect the company's profit.

☆ On the other hand, in family businesses, HRD was found to be proactive for family members as a part of medium to long-term development and succession strategies for designing future business plans. The small business owner needs to establish and maintain a productive working atmosphere for his or her work force. Employees only become a productive asset to the company when they feel that they are treated in a fair and professional way. The small business owner, who communicates business expectations and company goals to his employees, provides adequate compensation offers meaningful opportunities for career advancement and anticipates work force training and developmental needs have better chances to prosper in his business. These practices help in providing meaningful feedback to his or her employees too, which leads him to be more successful than that owner who is neglectful in any of these areas.

## Questions

### Short Answer Questions

1. Describe the features of HRM?
2. Explain Human Resource Development?
3. What is outsourcing of Human Resource?
5. Explain Academic Theory and Critical Theory of HRM?
6. What are the new HR Roles?
7. Explain Recruitment & Selection?
8. Explain Training and Performance Appraisal?
9. Describe the role of HR's Agenda in company's growth?
10. What is Growth gap?

### Long Answer Questions

1. Illustrate modern concept of human resources?
2. Discuss about Human Resource Management in detail?
3. Do you think that HR Manager's role is changing? Explain
4. Do you think that HR has to play a major role in organisational growth? If so, why?
5. Discus in detail about HR in small business.
6. Explain four roles of HR.

# Unit 4

# Career Planning, Training and Development

## 4.1 Introduction

If an organisation fails to procure employees with required qualifications, skill and caliber, a time may come when all the eligible employees will retire and the organisation is bound to suffer. Therefore, the importance of recruitment and selection of the right type of persons at the right time is indispensable to the organization's success.

Selection, either internal or external, is a deliberate effort of the organisation to select a fixed number of personnel from a large number of applicants. The primary aim of employee selection is to choose those persons who are most likely to perform their jobs with maximum effectiveness and remain with the company. Thus, in selection, an attempt is made to find a suitable candidate for the job. In doing so, naturally many applicants are rejected. This makes selection a negative function. In contrast, recruitment is a positive function because it attempts to increase the number of applicants applying for the job.

## 4.2 Career Planning

Career planning is structured exercise undertaken to identify one's objectives, marketable skills, strengths and weaknesses etc. as a part of one's career management. It is the process of establishing career objectives and determining appropriate educational and developmental programs to further develop the skills required to achieve short- or long-term career objectives. Career planning consists of activities and actions that you take to achieve your individual career goals.

Career planning is the ongoing process where you:

- Explore your interests and abilities
- Strategically plan your career goals
- Create your future work success by designing learning and action plans to help you achieve your goals.

From the organisational perspective, it refers to the development of forward looking HRM policies that creates an alignment between the expectations of employees and the organisation. Following are some common terminologies related to career planning:

- **Career Goal:** This refers to the aspirations with which an employee starts his career. It is the driving factor while choosing jobs.
- **Career Planning:** The process of selecting the career goals and the path to achieve the goals.
- **Career Path:** It is the sequential and progressive path through which one moves towards the selected career.
- **Career Anchors:** These are the basic drives acquired by an individual during the learning process that forms the basis for selecting a type of career.
- **Career Management:** It is step by step process of setting personal goals, selecting suitable career paths and preparing plan of actions to achieve the goals.

## 4.3 Significance of Career Planning

- Attract competent person having required abilities.
- Managing employees progression through promotions.
- Motivate employees for long term association.
- Identifying employees with potential.
- Optimum utilization of the organisational talent.
- Creating a learning competitive culture within organisation.
- Reducing employee dissatisfaction.

## 4.4 Recruitment

Recruitment refers to the process of attracting, screening and selecting qualified people for a job. For some components of the recruitment process, mid- and large-size organisations often retain professional recruiters or outsource some of the process to recruitment agencies. Recruitment is defined as a process of searching prospective employees and stimulating them to apply for jobs in the organisation. Recruitment is defined as, "a process to discover the sources of manpower to meet the requirements of the staffing schedule and to employ effective measures for attracting that manpower in adequate numbers to facilitate effective selection of an efficient workforce."

Recruitment of candidates is the function preceding the selection, which helps create a pool of prospective employees for the organisation so that the management can select the right candidate for the right job from this pool. The main objective of the recruitment process is to expedite the selection process. Recruitment is a continuous process whereby the firm attempts to develop a pool of qualified applicants for the future human resources needs even though specific vacancies do not exist. Usually, the recruitment process starts when a manger initiates an employee requisition for a specific vacancy or an anticipated vacancy.

**Fig. 4.1: Recruitment**

Recruitment needs are of three types:

- Planned *i.e.* the needs arising from changes in organisation and retirement policy.
- Anticipated Anticipated needs are those movements in personnel, which an organisation can predict by studying trends in internal and external environment.
- Unexpected Resignation, deaths, accidents, illness give rise to unexpected needs.

As seen, recruitment forms the first stage in the process which continues with selection and ceases with the placement of the candidate. Usually, the selection process starts with the indent for recruitment by the departmental heads. These indents specify the reasons why recruitment is to be made. These indents are sent to the personnel department. The personnel department has to check the financial implications of the recruitment to find out whether the additional expenses would be within the budgetary provisions. If everything is as per norms, the recruitment is allowed and the initial pay and other allowances are determined. Recruitment makes it possible to acquire the number and type of people necessary to ensure the continued operation of the organisation.

Also, recruitment is a process of searching for prospective employees and stimulating and encouraging them to apply for jobs in an organisation. Further, recruiting is the discovering of potential applicants for actual or anticipated organisational vacancies. Yoder is of the opinion that, "Recruitment is a process to discover the sources of manpower to meet the requirements of the staffing schedule and to employ effective measures for attracting that manpower in adequate numbers to facilitate effective selection of an efficient working force".

Thus, recruitment is the generating of applications or applicants for specific positions. It is a linking activity bringing together those with jobs and those seeking jobs. Recruitment is therefore the process of searching prospective workers and stimulating them to apply for jobs in the organisation.

### 4.4.1 Objectives of Recruitment

The objective of the recruitment process is to obtain the number and quality of employees that can be selected in order to help the organisation to achieve its goals and objectives. With the same objective, recruitment helps to create a pool of prospective employees for the organisation so that the management can select the right candidate for the right job from this pool.

Recruitment acts as a link between the employers and the job seekers and ensures the placement of right candidate at the right place at the right time. Using and following the right recruitment processes can facilitate the selection of the best candidates for the organisation.

- To attract people with multi-dimensional skills and experiences that suits the present and future organisational strategies.
- To induct outsiders with a new perspective to lead the company.
- To develop an organisational culture that attracts competent people to the company.
- To search or head hunt/head people whose skills fit the company's values.
- To devise methodologies for assessing psychological traits.
- To seek out non-conventional development grounds of talents.
- To search for talent globally and not just within the company.

## 4.5 Source of Recruitment Selection

Recruitments are a time taking process that includes many stages like looking for short listing etc. Moreover it is one of the prime functions as it involves adding new employees in the organisation. Not only time, recruitment also involves high costs. This means that it has to be managed with rigorous planning.

Source s of recruitments refers to the alternatives from where we can find the suitable candidates for filling the vacant positions. These vary depending upon the:

- Nature of job
- Levels of job position in the hierarchies like lower, middle or top management.

- ✰ Budget of the HR department.
- ✰ Number of vacancies to be filled.

Based on the above criteria the recruitment sources can be classified into two parts. These are:

1. Internal sources of recruitment
2. External sources of recruitment.

## 1. Internal Sources

Internal sources include alternatives of filling the position from within the organisation. It includes employees who are already working in the organisation. Among the internal sources, the following may be included:

### (a) Promotions

This is the most preferred source and refers to filling the position by promoting the employees from within. Based on the nature and level of job, suitable employees are promoted. This is a good way of motivating the employee. Moreover it provides reliable employees at minimum cost. However, there can be problems of merit, freshness and competition.

### (b) Transfers

This involves shifting the job location of an already working employee based on the vacancy. This usually occurs in manufacturing units, sales and marketing jobs. Transfers help in maximum utilization of the employee s experience and capabilities.

### (c) Response of Employees to Notified Vacancy

This is a method in which an opening is advertised within the organisation through different communication mediums so that employees who match the requirements can apply for the position. This method requires the employees to cross various stages of selection as per the prescribed process. This process provides opportunity to employees to change their jobs based on their interests and skills.

### (d) Databank

### (e) References

Internal sources of recruitments have both positive and negative effects. Though it is reliable source of recruitment, it can lead to interpersonal conflicts as well. Employers have to diligently plan the internal recruitment process to avoid any clashes related to seniority, reporting and team work.

## 2. External Sources

External sources refer to finding candidates for job openings from outside. Following are the sources for external recruitments:

### (a) Advertisements

This refers to informing the prospective candidates about the job opening through advertisements. Ads can be placed in different Medias like print, electronic, internet. The commonly used mediums are newspapers, job portals, and company website.

Advertisements are useful when the number of openings is large and company needs diverse workforce. This source is a good choice to fill the lower and middle level positions.

### (b) Campus Interviews

Incase the requirements is for fresher with less experience, campus selection can be used to find the candidates. Companies looking for professionals in specialized streams like Finance, ITI, HR, marketing, etc. can approach various professional institutions. Campus interviews help in getting the desired candidates at one place. The process of selection is also short and it's economical compared to other sources.

### (c) Recruitments Consultants

These are specialized professionals who undertake the recruitment process for various companies by charging fees usually based on percentage basis. Recruitments consultants use their extensive experience, database to select the appropriate candidates. It is useful in fulfilling the middle as well as top level vacancies. Companies outsource their recruitment activities to these professionals.

### (d) Employment Agencies

These are government owned employment exchanges that fill mainly the lower level jobs for skilled and semi-skilled workers.

### (e) Job portals and Internet

Recruitments process has become easy as internet has opened various options for the employers. The various methods of recruiting on internet are:

- ☆ **Job portals:** Job portals are specialised internet platforms for searching jobs as well as candidates. Some of the well-known job portals are Naukri.co, monster.com, shine .com etc. These sites contain the vast data of jobseekers who are registered user. Job portals help in finding the most suitable candidates who are seriously looking for a job.
- ☆ Social networking sites like LinkedIn, Facebook and Google+ etc. are open platforms where employers can find their prospective employees. These are informal channels of recruitments.
- ☆ **Video conferencing:** Globalization has opened many new sources like video conferencing. Employers save time and cost by conducting interviews on internet through camera.

### (f) Direct Recruitment

Usually for workers and staff this includes placing an advertisement for recruitments on the company ate. Like walk-in etc. usually labour class people keep looking for jobs as daily basis. Companies also find this a suitable option to fill the contractual and casual labours at lowest cost.

## 4.6 Selection

Having identified the potential applicants, the next step is to evaluate their experience and qualifications and make a selection. Selection can be conceptualised in terms of either choosing the fit candidates or rejecting the unfit candidates or a combination of both. According to Yoder, "the hiring process is of one or many „go-no-go-gauges. Candidates are screened; the applicants go on to the next hurdle, while unqualified candidates are eliminated".

Great attention has to be paid to selection because it means establishing the "best fit between job requirements on one hand and the candidates qualifications on the other". Faulty judgment can have a far-reaching impact on the organisational functioning. There are several advantages of a proper selection procedure, as the employees are placed in the jobs for which they are best suited. They derive maximum job satisfaction, labour turnover is reduced and the overall efficiency of the concern is increased. And finally, a good relationship develops between the employer and his workers.

Therefore, in simple terms, selection is a process in employment function which starts immediately upon receipt of resumes and application letters, the major concern being reviewing resumes for basic qualifications. A job seeker who does not meet the required qualifications is not an applicant and should not be considered. It is a process which should be based on job-related qualifications, including but not limited to: required or preferred education; experience; and knowledge, skills and abilities as identified in the job description.

Qualifications must be bona fide occupational qualifications. An applicant who is hired must meet the required qualifications listed in the job description. In this regard, selection is a process of matching the qualifications of applicants with the job requirements. It is a process of weeding out unsuitable candidates and finally identifies the most suitable candidate.

## 4.7 Steps of Selection Procedure

At the end of the recruitment process, HRM department receives resumes of interested candidates. The next step is selecting the most suitable candidate from these applications.

Selection process deals with evaluating all the profiles received through recruitment drive. The process involves in-depth assessment of candidates qualification, experience, skills, behaviors and attitudes based on the requirement of the vacant positions.

An ideal selection process fulfills the following criteria:

- Accumulating information of the employer and prospective employee like employment conditions, personal characteristics, career paths, experience, qualification etc.

- Prediction: The information of candidate's past and present job helps in predicting his future course of actions. These prediction are helpful while selection.
- Decision Making: The predictions made above help in decision making for finals selection.

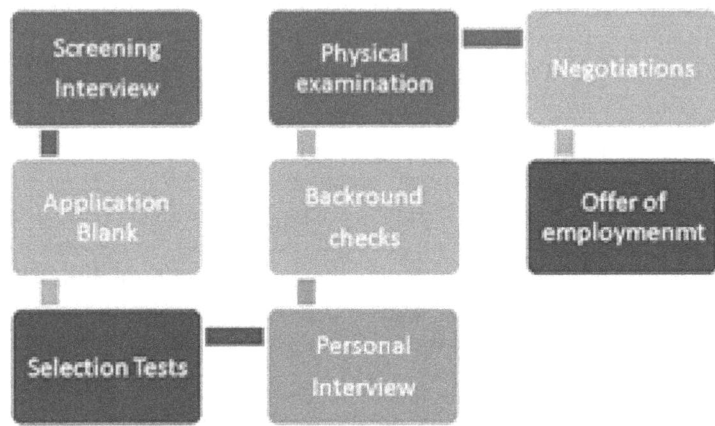


**Fig. 4.2: Steps in Selection process**

An ideal selection process therefore consists of the following eight Steps:

1. **Screening Interview:** This is the first step in which HR professionals shortlist the eligible candidates after matching their profile with Job specifications and Job descriptions. The shortlisted candidates are then interviewed, usually over telephone to find out their interest in job, expectation and availability for selection process.
2. **Application Blank:** The candidates shortlisted through the telephonic interviews are asked to fill a company generated application also called application blank. This blank is an important document of personal records with the company. An application blanks usually contains the following data:
   - Personal information like age, marital status, achievements, competencies and nationality etc.
   - Qualifications both educational and professional.
   - Work experience including organisation name, period of work, salary, responsibilities etc.
   - References from previous employers
   - Extracurricular information related to sports, arts, etc.

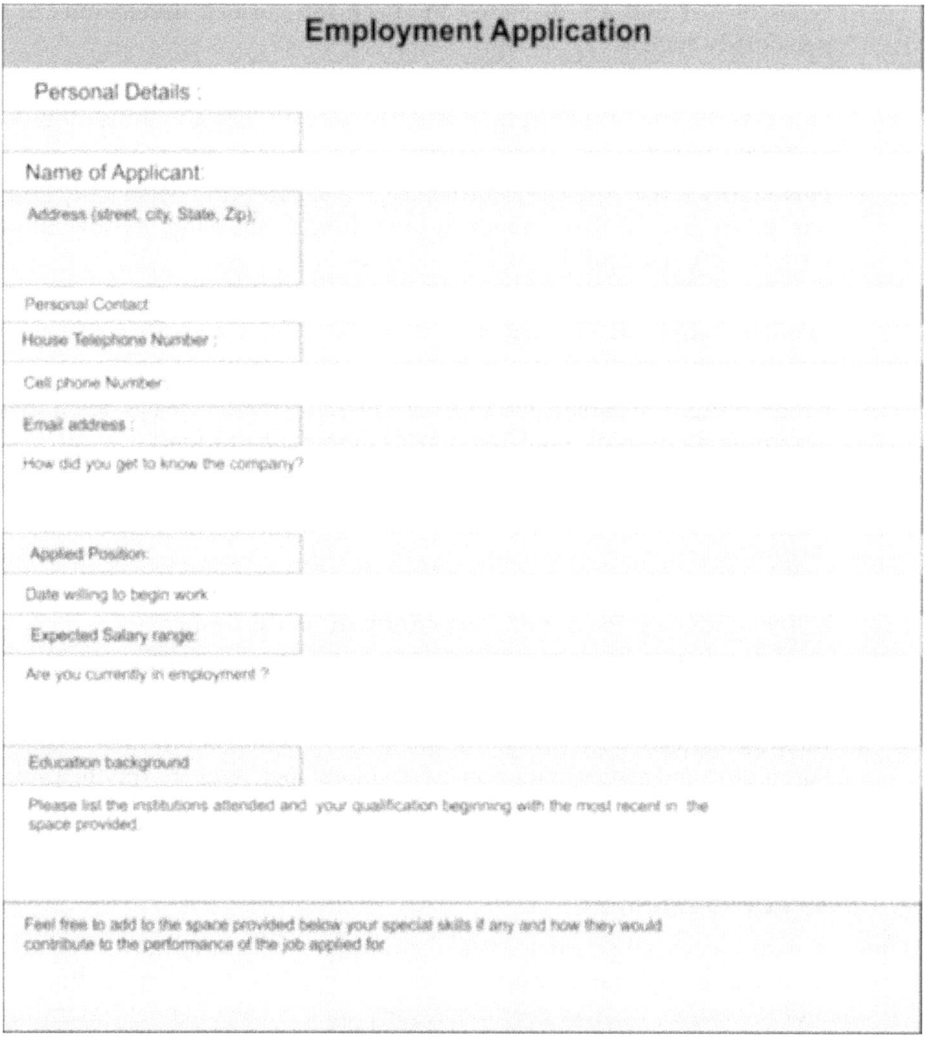

**Employment Application**

Personal Details :

Name of Applicant:

Address (street, city, State, Zip):

Personal Contact

House Telephone Number :

Cell phone Number:

Email address :

How did you get to know the company?

Applied Position:

Date willing to begin work :

Expected Salary range:

Are you currently in employment ?

Education background

Please list the institutions attended and your qualification beginning with the most recent in the space provided

Feel free to add to the space provided below your special skills if any and how they would contribute to the performance of the job applied for

**Fig. 4.3: Employee Application**

## 3. Selection Tests

The next step requires candidates to pass the selection tests based on the job. According to Lee J. Groombach," Test is a systematic procedure for comparing the behaviors, skill and abilities of two or more persons". The main objective of conducting selection test is to find confirm the information provided in the resume. Tests also help in finding that information that is unknown through application blank.

There are broadly many types of tests:

☆ **Psychological tests:** These tests test the soft skills like behaviors, abilities, stress, personality, team work, emotional intelligence etc. There are many types of test that are designed to check the various dimensions of a personality such as:

- **Projective tests:** These are picture based tests used to assess the perception of a person. The way he responds to the picture tells about his attitude, apprehensions etc.
- **Personality tests:** These tests are conducted to assess the personality traits like interpersonal competence, dominance, submission, extroversion, introversion, self-confidence, leadership etc.
- **Polygraph tests:** These are also known as lie detector tests and are used in case job is vulnerable to theft etc.
- **Graphology:** These are assessing an individual on the basis of their handwriting. The personality's type is determined on the basis of the style of writing for example big letter depicts dominant attitude.
- **Technical tests:** These test are designed to test the technical competencies of the candidates like computer knowledge, typing skills, etc. types of test include:
- **Aptitude Tests:** These tests assess the skill, knowledge of a person. The test help in find out if the person can perform the job as desired.
- **Achievement tests:** These are tests in fact the scores of previous tests that portray a candidates eligibility.
- **Intelligence tests:** Such tests aim at assessing the mental ability of candidates. For middle and senior management positions such tests are very helpful.
- **Judgment tests:** Also called problem solving tests, in these candidates are provided with real business situation and asked to provide solutions.

## 4. Physical Examination

After successfully crossing various tests, the next stage is to appear for physical examination. This is important to assess a candidate's ability to perform in different job situations, location, weather conditions, physical labor, eyesight etc. Many big organisations have their own doctors who conduct these tests. Such examination can be found in Indian army employment tests, Banks, Police etc.

## 5. Back Ground Checks

The next step is reference checking background of an employee. Also called as reference checks this step is of utmost important to confirm the nationality, experience and qualification of the candidate. There are many specialized agencies that work full time to provide such services to employers.

## 6. Personal Interview

Candidate who successfully passes all the above stages is called to appear for a final interview. This is the first face to face interview of the candidate with the management. The interview is usually taken by panel of experts and consists of 2-3 rounds. The rounds are also called as:

- HR round
- Technical round
- Final round

## 7. Negotiations

The candidate who clears the final sound is called to discuss on salary, job profile, working conditions, perks and benefits. The management puts their conditions and the candidate is asked to tell his expectation. When both match, the candidate is finally selected.

## 8. Offer of Employment

One the terms and conditions are finalized between the employee and management, the next immediate step is to present the offer letter.

### Appointment

Once the selection decision has been made, the notification of consideration has been issued and the waiting period is over, only then can the offer of appointment be made.

An oral offer of appointment can be made first followed by a written offer that clearly sets out all of the conditions of the appointment. Any offer, whether oral or written, must be made by a person with the sub-delegated authority to do so. It is important to remember that the offer must be made in the official language of the person's choice and be provided in accessible formats, if required.

### Conditional Offers

In certain situations, when it is not possible to make a firm offer of employment, a conditional letter of offer may be issued. A conditional offer can only be made by a person with the sub-delegated authority. It is as legally binding as an offer without conditions, subject to the conditions being met. There may be a number of reasons why a conditional offer is needed, (e.g., to maintain the interest of prospective employees in a competitive marketplace or to help speed up the appointment process by allowing persons to give notice from their current employment). It should be made clear to recipients of conditional offers that if the condition is not met, the offer is no longer valid.

An offer letter is a formal intimation to the prospective employees of selection for a specific post and essentially includes:

- Confirmation of offer for employment in the organisation
- Tentative Date of joining
- Salary particulars(Optional)
- Location of job
- Job designation

It would not be wise for the person to quit their current job until the offer is final. The following are examples of situations where a conditional offer might be issued:

- ✰ When offering an appointment to a student who applied on an internal advertised appointment process, but who has NOT yet completed their studies, an offer conditional on the person providing proof of completion prior to the effective date of the appointment could be made.
- ✰ While awaiting confirmation that the person meets a condition of employment, such as a security clearance, on the condition that clearance is granted prior to the date of appointment.

Name and Address

Date

Dear Mr./Miss/Mrs./Ms. (Name):

Congratulations! We are pleased to confirm you have been selected to work for (Company/Division/Department). We are delighted to make you the following job offer.

The position we are offering is that of (Job Title) at a salary/wage of (salary/hourly rate) per (year/month/week/hour). This position reports to (Title and name of Supervisor). Your working hours will be from (state working hours) and/or (normal workdays). This is a (permanent/seasonal/contract/casual) position. (If this is a contract position state expected length of term).

Name and Address

Date

Dear Mr./Miss/Mrs./Ms. (Name):

Congratulations! We are pleased to confirm you have been selected to work for (Company/Division/Department). We are delighted to make you the following job offer.

The position we are offering is that of (Job Title) at a salary/wage of (salary/hourly rate) per (year/month/week/hour). This position reports to (Title and name of Supervisor). Your working hours will be from (state working hours) and/or (normal workdays). This is a (permanent/seasonal/contract/casual) position. (If this is a contract position state expected length of term).

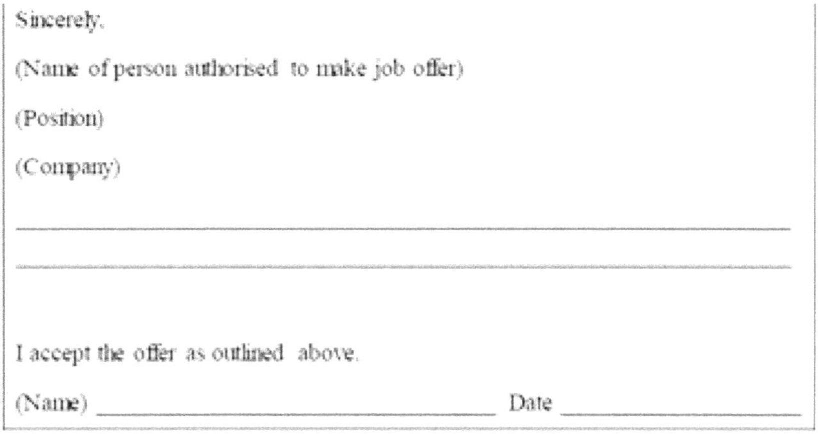

Sincerely,

(Name of person authorised to make job offer)

(Position)

(Company)

I accept the offer as outlined above.

(Name) ____________________ Date ____________

**Fig. 4.4: Sample**

Appointment letter also known and employment contract is a formal document setting the terms and conditions for the relation between the management and the employees. The T&C are legally binding and both parties are required to give their consent by signing the documents. An appointment letter is also important as it sets the guidelines for work activities, employer's expectation and company rules and policies.

The points covered in an appointment letters are related to:

- Job details like designation, location and grade
- Salary and benefits including the benefits, rewards, incentives etc.
- Reporting structure
- Transfer clauses
- Resignation clauses
- Confidentiality clause
- Working conditions
- Other Miscellaneous clauses varying depending on the organisation policy.

## 4.8 Training and Development

Organisations and individuals should develop and progress simultaneously for their survival and attainment of mutual goals. Employee training is a specialized function and is one of the fundamental operative functions of human resource management.

Training improves changes and molds the employee's knowledge, skill, behaviour, aptitude and attitude towards the requirements of the job and the organisation. Training bridges the difference between job requirements and employee's present specifications.

Training and development is a subsystem of an organization which ensures that randomness is reduced and learning or behavioural change takes place in structured format.

### 4.8.1 Traditional and Modern Approach of Training and Development

Traditional Approach is based on the fact that most of the organisations before never used to believe in training. They were holding the traditional view that managers are born and not made. There were also some views that training is a very costly affair and not worth. Organisations used to believe more in executive pinching. But now the scenario seems to be changing.

The modern approach of training and development is that Indian organisations have realized the importance of corporate training. Training is now considered as more of retention tool than a cost. The training system in Indian Industry has been changed to create a smarter workforce and yield the best results.

### 4.8.2 Objective of Training and Development

The principal objective of training and development division is to make sure the availability of a skilled and willing workforce to an organisation. In addition to that, there are four other objectives: Individual, Organisational, Functional and Societal.

- ✰ Individual objectives help employees in achieving their personal goals, which in turn, enhances the individual contribution to an organisation.
- ✰ Organisational objective assist the organisation with its primary objective by bringing individual effectiveness.
- ✰ Functional objectives maintain the departments contribution at a level suitable to the organisations needs.
- ✰ Societal objectives ensure that an organisation is ethically and socially responsible to the needs and challenges of the society.

### 4.8.3 Procedure

Training programme is a costly and time-consuming process. The training procedure discussed below is essentially an adoption of the job instruction-training course. The following steps are usually considered as necessary.

1. Discovering or identifying training needs
2. Preparing the instructor or getting ready for the job
3. Preparing the trainee
4. Presenting the operation
5. Try out the trainees performance
6. Follow-up or rewards and feedback

#### 1. Discovering or Identifying the Training Needs

A training programme should be established only when it is felt that it would assist in the solution of specific problems. Identification of training needs must contain three types of analysis:

(a) Organisational Analysis: Determine the organisation s goals, its resources and the allocation of the resources as they relate to the organisational goals

(b) Operations analysis focuses on the task or job regardless of the employee doing the job.

(c) Man analysis reviews the knowledge, attitudes and skills in a person that must be acquired to contribute satisfactorily to the attainment of organisational objectives.

Armed with the knowledge of each trainee's specific training needs, programmes of improvement can be developed that are tailored to these needs. The training programme then follows a general sequence aimed at supplying the trainee with the opportunity to develop his skills and abilities.

### 2. Preparing the Instructor

The instructor is the key figure in the entire programme. He must know both the job to be taught and how to teach it. The job must be divided into logical parts so that each can be taught at a proper time without the trainee losing perspective of the whole. This becomes a lesson plan. For each part one should have in mind the desired technique of instruction, *i.e.*, whether a particular point is best taught by illustration, demonstration or explanation.

### 3. Preparing the Trainee

This step consists of:

(a) Putting the learner at ease

(b) Stating the importance and ingredients of the job and its relationship to work flow

(c) Explaining why he is being taught

(d) Creating interest and encouraging questions, finding out what the learner already knows about his job or other jobs

(e) Explaining the „why of the whole job and relating it to some job the worker already knows

(f) Placing the learner as close to his normal position as possible

(g) Familiarizing him with the equipment, materials, tools and trade terms

### 4. Presenting the Operations

This is the most important step in a training programme. The trainer should clearly tell, show, illustrate and question in order to put across the new knowledge and operations. There are various alternative ways of presenting the operation namely, explanation and demonstration etc. An instructor mostly uses the method of explanation. In addition one may illustrate various points through the use of pictures, charts, diagrams and other training aids. Demonstration is an excellent device when the job is essentially physical in nature. The following sequence of training may be followed:

(a) Explain the sequence of the entire job.

(b) Do the job step by step according to the procedure.

(c) Explain each step that he is performing.

(d) Have the trainee explain the entire job.

Instructions should be given clearly, completely and patiently; there should be an emphasis on key points and one point should be explained at a time. The trainee should also be encouraged to ask questions in order to indicate that he really knows and understands the job.

#### 5. Try out the trainee's performance

Under this, the trainee is asked to go through the job several times slowly, explaining each step. Mistakes are corrected and if necessary, some complicated steps are done for the trainee the first time. Then the trainee is asked to do the job, gradually building up skill and speed. As soon as the trainee demonstrates that he can do the job in the right way, he is put on his own. The trainee, through repetitive practice, will acquire more skill.

#### 6. Follow-up

The final step in most training procedures is that of follow up. This step is undertaken with a view to testing the effectiveness of training efforts. The follow up system should provide feedback on training effectiveness and on total value of training system.

## 4.9 Evaluation of Training and Development Programme

Training and development is an important part of assisting performance improvement at organisational, faculty/central department, unit and individual levels. It is therefore important that the transfer of learning into the workplace is assessed through a process of review and evaluation so that its success or otherwise can be established and so that we can demonstrate the contribution learning makes towards overall organisational success. Evaluation is the process of finding out how the development or training process has affected the individual, team and the organisation.

Evaluation methods can be either qualitative (e.g. interviews, case studies, focus groups) or quantitative (e.g. surveys, experiments). Training evaluation usually includes a combination of these methods and reframes our thinking about evaluation in that measurements are aimed at different levels of a system.

### 4.9.1 Benefits of evaluating training and development programme

The benefits of evaluating training and development are to:

- Promote business efficiency by linking efforts to train and develop staff to operational priorities, goals and targets
- Identify cost effective and valuable training events or programmes, leading to better focused learning and development
- Ensure the transfer of learning into the workplace
- Use and reinforce techniques learned to help improve quality and customer service within the organisation

- Help define future development objectives

## Techniques of Evaluation

Various methods of training evaluation are:

- Observation
- Questionnaire
- Interview
- Self-diaries
- Self-recording of specific incidents

### 4.9.2 Stages of Evaluation

There are four key stages at which training and development should be evaluated:

1. **Reaction:** At this stage, evaluation provides information on the attitudes and opinions of participants to the learning they have undertaken typically via evaluation forms or comment sheets. It provides useful information to allow assist with modifying the curriculum/training programme. What did the participant think of the development activity?
2. **Learning attained:** Evaluation at this stage looks at the extent to which learning objectives have been achieved. Evaluation of learning can take place during the activity using interactive sessions, tests and practical application. Did the participant learn what was intended? Were the learning objectives met?
3. **Performance:** Evaluation at this stage looks at the impact of a learning experience on individual/team performance at work. Key to this level of evaluation is the need to have agreed clear learning objectives prior to the learning experience so that when evaluation takes place, there are measures to use. Did the learning transfer to the job? How has the development activity improved individual performance, for example specialist knowledge or professional approach?
4. **Organisational Impact:** At this level evaluation assesses the impact of learning on organisational effectiveness and whether or not it is cost effective in organisational terms. How has this development activity affected the organisation, faculty, central department or unit in terms of improved performance – for example, better results, enhanced quality or standards, financial stability, fewer complaints, increased morale, professional image?

## Questions

### Short Answer Questions

1. Differentiate between recruitment and selection.
2. Explain the concept of selection in Yoder s view.
3. What do understand by initial and main interviews?

4. Explain employment test.
5. What is an appointment letter? State the various elements of an appointment letter.

## Long Answer Questions

1. Explain the techniques and stages of evaluating training and development.
2. Discuss the various methods of training and development and comment on the benefits of evaluation of training programs.
3. What is career planning? Describe the elements of career planning.

# Glossary

**Behaviour Modeling :** It is a training technique in which trainees are first shown good management techniques in a film, and then asked to conduct role play in a simulated situation. Afterwards they are given feedback and praise by their superior.

**Man Analysis Reviews :** It is an analysis if personal qualities, traits, skills and competencies of individual employees. The process helps in ascertaining training needs.

**Career Anchors :** Refers to the attitudinal syndromes that develop during the socialization process. These are the choice of careers based on personal need and talent that decides the career development. Career anchors are those choices that an individual do not want to part with.

**Offer Letter :** It is the formal letter being issues by the employer on finally electing a candidate in the selection process. The letter consists of basic employment details like position, salary and date of acceptance of offer.

**Reference Checks :** It is the cross examination of a shortlisted candidate during the selection process. It checks the background and details provided by him to find out the authenticity.

# Unit 5

# Agri Business Planning

## 5.1 Introduction

Future outcomes are a function of today's decisions. Although there is a high degree of randomness and uncertainty associated with the future, you can increase the probability of a successful outcome by planning ahead. This is true in nearly every aspect of our lives, both personal and professional. For those who operate their own businesses, planning becomes increasingly important because the personal and professional aspects become more difficult to untangle. In agricultural businesses, planning may be even more vital because of the inherent uncertainty associated with agricultural production. Some important sources of uncertainty include production risk, price risk, financial (or interest rate) risk, and changes in government programs.

## 5.2 Business Plan

One of the most important documents for any business is their business plan. It is common practice for consultants, lenders, potential business partners, and other business-associated individuals to request a business plan to make a more informed decision concerning their relationship with a business. However, business plans have many more direct benefits for the business owner. The planning process forces owners to systematically consider all facets of the business. In so doing, they become more knowledgeable of the business, the industry, and the market environment in which their business operates. The process also helps to define business goals and to assess the impact that uncertainty may have on future business outcomes. Perhaps most importantly, the written plan provides a well-defined direction for the business. Therefore, it can be used to keep all employees moving toward the common goals established within it.

Completing a business plan can be a time-consuming activity, but well worth the effort. Because businesses operate in an ever-changing environment, the plan should be revisited periodically to be sure that the business is headed in the proper direction or to formally alter the firm's course if circumstances dictate that this is necessary. Again, the systematic review of the business plan forces the owner, and potentially others, to look at the business as a whole and make better-informed decisions.

### 5.2.1 Writing an Agri-business Plan

A good business plan gives credibility to your agri-business, and to your skills as a agri-business owner or manager. Even if you do not intend to make any changes to your agri-business, you should still write a business plan. A good agri-business plan can highlight weaknesses in how you plan and run your agri-business, which can provide helpful insight for later improvements.

Before writing a business plan, it is advisable to thoroughly research the sectors that you are currently working in - or plan to work in - including any future economic prospects for these sectors. You should also familiarise yourself with the relevant regulations for your business.

When writing your business plan, consider including:

- ☆ Your short-term and long-term aims
- ☆ A timescale for achieving your aims
- ☆ Who will be involved
- ☆ How you will manage the money

It may also describe your agri-business's unique characteristics, for example, its:

- ☆ Location
- ☆ Soil type
- ☆ Management eg. whether it is family run

### 5.2.2 Business plan structure

Most business plans contain:

- ☆ Financial forecasts
- ☆ Your marketing and sales strategy
- ☆ Information about your management team and staff
- ☆ An operations plan

Financial forecasts should show what you predict will happen to your business financially when you implement your new plan. For example, you should think about:

- ☆ How much external funding you may need

- What you can offer as security against loans - if needed
- How much income you are expecting
- What your expected profits and losses will be
- Your financial forecasts should include a cash flow forecast, and projected profit and loss account.

The operations plan is a description of the agri-business itself, and how it is run. It can include details about:

- The land, buildings and facilities
- Equipment, vehicles and machinery used on the agri-business
- Materials and supplies
- What is produced and when
- Plans for new facilities
- Any environmental assessment plans, eg for soil conservation or to improve water quality
- Relevant regulations and licences

It may also be useful to prepare an executive summary - a synopsis of key points from your entire plan and a short description of your business opportunity - for presenting to third parties such as your bank, potential investors or suppliers.

## 5.3 SWOT Analysis

Performing a SWOT analysis, which stands for strengths, weaknesses, opportunities, and threats, lays the foundation for the business plan. Four separate SWOT analyses should be performed, each related to one of the four functional areas of management: marketing, production/operations, finances, and human resources.

When assessing strengths and weaknesses, the focus should be internal. Opportunities and threats, on the other hand, should reflect external factors. For example, proximity to a major market, such as a large city, may provide an opportunity to market processed dairy products directly to a restaurant. Threats may take the form of new competitors or changes in agricultural production or environmental policy.

### 5.3.1 Performing SWOT Analyses

Simply divide a piece of paper into four quarters, label the quarters appropriately, and begin to write your thoughts down.

- Business Plan Format
- Sections of Business Plan
- Introduction
- Marketing Management
- Production/Operations Management

- Human Resources Management
- Financial Management
- Summary
- Appendices

As noted, it covers the four managerial functional areas. We present the example structure and provide some ideas for what you will want to include in each section. Creating a thorough document the first time through is important. This will make follow-up revisions easier to implement.

## Introduction

The introductory section gives a broad overview and background of the business. Several subsections (outlined below) should be included to provide a thorough overview. However, if there's something that you feel isn't applicable to your business, feel free to omit it from your plan.

## Title Page

The first page should give the name of the document, the firm's name, and the names of all those involved in developing the plan. Dating the plan so that you can remember when it was developed or updated is also wise.

## Executive Summary

This section, while appearing at the front of the business plan, is actually the last piece developed. Here you should present the most important information, which may include the firm's goals and objectives and associated target dates. Basically, the executive summary provides a concise overview of the business plan.

## Table of Contents

The table of contents should provide the titles of all section headings in the plan and the page numbers on which the sections begin.

## Vision and Mission Statements

These relatively brief statements tell the reader why the business is in operation and where the management team, or owner(s), plans to be in the future. The vision statement should tell the reader what business the firm is in, or plans to enter, and what the most important business goals are. That is, it should tell where the firm is going.

## Overview of Productive Assets

Outline what productive assets are necessary to make your product or provide your service. The following deserve particular attention:

- Land
- Buildings
- Other Facilities (particularly if on-agri-business processing is involved)

- Equipment and Machinery
- Materials and Supplies

Focus on what the firm currently owns, the quality of those assets, and how others will be obtained, if needed. Only discuss the resources needed. Save any discussion of financing these assets for the Financial Management section.

## Managerial Expertise

Here again, note the special expertise held by management in the area of production/operations. For example, is the herdsman on a dairy agri-business trained in dairy science? Does the crops manager have a background in agronomy, soil science, or other related field? How many years of experience in this type of position does this person have?

## Production/Operations Performance

Describe your current production practices. How much do you produce? When do you produce it? You may want to develop a visual approach to help tell this story. A time line could help to describe when products are made or services are sold. The more complex the business is, the more useful a visual might be.

## Regulatory Considerations

Government regulations affect production in many industries. This is particularly the case in production agriculture. If producing agricultural products, be sure that you are complying with all relevant regulations. These include, but may not be limited to:

- Manure management
- Soil conservation
- Worker Safety
- Zoning
- Inspections of the product and of the production facilities

You can gain information on relevant policies and regulations from business consultants, Penn State Extension, agricultural cooperatives, and government agency Web sites.

## Production/Operations Strategy

Now that you have laid out your current production practices and defined some firm and industry trends as they relate to some important benchmark measures, it is time to describe your production- or operations-related plans. As you do this, be sure to set specific production goals, outline potential changes in enterprises or production practices, and describe how you plan to locate and purchase inputs.

- How long will your current productive assets be of use? When do leases on land and equipment expire? How soon can you expect to replace important machinery and equipment?
- Where can you find other inputs, such as feed, in the future—particularly if you are expanding your operation?

- ☆ Should you consider hiring a custom operator to perform a portion of the production tasks?
- ☆ Are there new production practices or machinery you should consider adopting?
- ☆ How many units of product do you want to sell over your planning period? Provide specific targets and a time line, if appropriate.
- ☆ Do you need to develop a nutrient management plan or update an existing one? Are there other environmental plans that should be developed?
- ☆ If expanding, how will new construction or other changes affect output? How will these changes affect your resources? Will you be able to operate in a timely manner without affecting the quality of your product(s)?
- ☆ Are there new enterprises that should be explored?

## Overview of Current Policies

Whether formalized or not, your firm has HR policies. While we encourage you to formalize those if they are not already, you should complete this section with the most accurate information you have at your disposal. Think about the following individual points as you consider what you would like to include in this section. Also, remember to include information regarding all employees of the business, including the management team.

Compensation and Benefits (Incorporation of this information may dictate that a portion of the plan be labeled as confidential. Thus, only certain members of the ownership or management team would have rights to view this type of information. you are trying to hide information.)—How much do you pay your employees? At what intervals are they paid (for example, weekly or monthly)? What sort of benefits package is offered? Does the package differ by type of employee? Do you have an incentive plan for employees? Are owner/ operators paid a salary or do they capture retained earnings?

1. **Job Descriptions and Recruiting :** Does each position have a written job description? Are these used to assess the suitability of potential employees? How do you recruit new workers/managers?
2. **Training and Standard Operating Procedures :** What training is provided for new employees? What training is provided when employees assume new responsibilities? Are common task sequences documented with written standard operating procedures (SOPs)?
3. **Performance Evaluation and Performance Feedback :** Is there a formal mechanism for evaluating workers' performances? If so, how frequently is performance assessed? How does the employee receive the manager's assessment? Are salaries or wages based at least partially on these evaluations?

## Managerial Expertise

Business provides materials to help develop job descriptions, SOPs, and other important HR documents. Business consultants with expertise in HR management should also be used in many instances to help the business owner develop the best HR policies possible for that particular business.

## Human Resource Strategy

Once you have outlined your production and marketing plans, you must evaluate the ramifications of those plans for the firm's human resources. Will the plan require any shift in HR policy? If so, how? For example, an expanding custom heifer grower may need to hire a full-time nutritionist to be sure that the heifers are receiving a properly balanced ration as they develop. The best way to reflect these changes may be to provide one or more additional organizational charts showing how the organization is expected to change over time. In the text, be specific as to what changes are to be made and when.

Also, be sure to describe any changes you plan to make in your HR management. If you don't have formal job descriptions, standard operating procedures, or evaluations, for example, you should consider putting those in writing. Also indicate if the plan will require additional training of existing and new employees. Finally, if expanding, describe where you might find potential employees.

## Financial Management

This section is the most crucial from a potential lender's perspective. Here, you should tie together the details in the rest of the plan in terms of how they affect the firm's financial performance. Ultimately, operating a business is about making money. Therefore, this section needs to allow the reader to assess where the firm is and where it intends to go over the planning horizon. Although you should provide current and projected future ("pro forma") financial statements with the plan, they might be best presented in an appendix. This section should be mostly a verbal explanation of the business's finances, with perhaps a few tables to highlight important information.

Because this section is so important, especially if financing is being pursued, we highly recommend that you work with a business consultant, accountant, or other financial advisor to develop it.

## Financial SWOT Analysis

Perform a SWOT analysis of the firm's financial position. Unlike some other areas, your frontline workers may not have as much to provide in this analysis. However, depending on the firm's "culture," (the accepted values and norms under which it operates. Some firms may be quite "laid back," allowing employees a good bit of decision-making authority. Others might be more "straightlaced" following a well-established set of rules, whether written or unwritten) you may still want to invite their input.

## Review of Current Financial Situation

Here you should highlight the important points of your financial statements (income statement, balance sheet, and statement of cash flows). Focus on the positive aspects, while not ignoring the negative. You do not want to provide a potential lender with an impression that you are trying to hide information. It might help if you work with a financial advisor to develop this narrative.

Provide a table of current outstanding debt. Include the terms of the debt, the lender(s), the principle amount(s), your payment amount(s), how frequently you make payments, and how many payments remain. Furthermore, a table of financial ratios would be useful in providing a snapshot view of the firm.

## Financial Strategy

At this point, you need to set forth your plan for financing the firm's operations over the planning period. Where will you get money when you need to purchase a new truck or replace your barn, for example? Present the highlights of the pro forma financial statements as discussed earlier. Also, you should relate your financial plan to your production, marketing, and human resources plans. A time line relating events planned in the other sections to financing may help to clarify this section's message for the reader.

In your discussion, be sure to let the reader know how you will assess financial performance. As in other sections, be specific. Will you require that net income grow at 8 percent per year, for example? Set goals for the measures you have used previously to describe the health of the business. Therefore, you should aim for specific values for your selected measures of profitability, financial efficiency, solvency, and liquidity. Also, do not forget that an information management system should be in place so that the financial data you gather is accurate. (An information management system (IMS) is any system that you can use to track important information regarding financial performance, in this case. Actually, you should have a system of information management that provides high-quality information for each of the four facets of management. This might include a production record-keeping system (DHIA for dairy agri-businesss), an accounting system, an inventory list, employee time sheets, and so forth.)

You should be realistic with your plans, yet push yourself. Stated differently, you should plan to succeed, not just survive. If your business is to be viable over the long term, then you should generate returns to grow the business, grow equity, improve your credit-worthiness, and otherwise improve the odds of operating this business well into the future. If the plan covers a major shift in the business's operations, such as a large expansion, then special care is needed to discuss how cost overruns might be handled, when production will begin in a new facility, when debt repayment will commence, and so forth. Although the planning process should reduce the amount of uncertainty associated with such a change, it can never eliminate uncertainty. Therefore, it should be noted that insurance may be used to protect the firm against financial losses that may be associated with operating a business. Be sure to define your insurance needs in this section of the business plan.

Uncertainty should also be accounted for in your financial forecasts. Let the reader know what assumptions you have made when developing the proforma statements. Also, some sensitivity analyses would be useful to show how your statements would change if output or input prices were different from your projections. If appropriate, set forth some contingency plans to be enacted if certain undesired outcomes are realized. For example, if milk production suffers from a hot, dry summer, you should have a contingency plan in place to help cash flow the business until production increases. For example, such a plan might include a revolving line of credit with the local bank.

## Questions

### Short Answers

1. What is Financial Management?
2. What is agri-business?
3. Explain Human Resource Strategy?

### Long Answers

1. How to write an Agri-business Plan?
2. Explain about SWOT analysis?

# Glossary

**Asset Turnover Ratio:** The percentage of total assets earned as gross income. A higher number is generally associated with higher profits. Mathematically, this equals gross income divided by average total productive assets.

**Balance Sheet:** A financial statement that shows total assets, total liabilities, and owners' equity at a specific point in time. The liabilities and owners' equity represent claims on the firm's assets.

**Business Planning:** The process of analyzing the firm's strengths, weaknesses, opportunities, and threats, using that information to develop organizational goals, and crafting strategies to reach those goals.

**Competitive Advantages:** Refers to particular strengths of the firm relative to those of other firms. Some examples are being the first firm in an area to provide custom heifer raising, having a manager with strong direct-marketing skills, or having superior land for growing crops.

**Contingency Plans:** These are strategies for dealing with potential outcomes that differ from those assumed in the initial planning process. These are most frequently associated with unfavorable outcomes.

**Contract Production:** Refers to any situation in which the farmers grows crops or livestock for a specific firm under terms negotiated in a contract.

**Extension:** The system of offices located around individual states that provides information and education to agri-business managers and other individuals.

**Current Ratio:** Measures the ability of the firm to pay its current liabilities with its current assets. A ratio greater than one indicates that the firm is liquid and able to cover its current liabilities. Mathematically, this is equal to current assets divided by current liabilities. Current assets include cash and other assets that will be converted to cash or used up within one year. Current liabilities are those that are payable within one year.

**Custom Business:** Any firm offering to perform services for a agri-business that would replace those already provided by the agri-business's labor. Some examples include crop scouting services, custom planting and harvesting, or custom heifer growing.

**Debt to Asset Ratio:** Indicates the percentage of total assets owned by creditors. For example, a debt to asset ratio of 0.5 means creditors own 50 percent of the agri-business's assets. Mathematically, this is total debt divided by total assets.

**Income Statement:** Provides a review of revenues and expenses over a given period of time, often a year. This may also be referred to as a profit and loss statement, earnings statement, or an operating statement.

**Leverage Ratio:** This represents total agri-business debt as a percentage of equity. If this ratio is greater than one, for example, then the business is financed by debt more than by equity. Mathematically, this is total debt divided by equity.

**Mission Statement:** Provides a summary of why the business is in operation. This may include the firm's common values, an overview of products or services, target markets, or other information to provide a clear picture of the firm's purpose.

**Net Income:** Represents the difference between gross income and total expenses. Mathematically, this equals gross income minus total expenses. A positive number means that the business is making enough to cover expenses and either reinvest in the company, pay debt more quickly, or increase owner incomes.

**Operating Expense Ratio:** Represents the percentage of gross income used in operating expenses (those expenses on inputs used in the current period). Mathematically, subtract interest expenses from operating expenses and divide the result by gross income.

**Organizational Chart:** A graphical representation of the formal chain of command for a firm. It shows who the supervisors are and over whom these have authority.

**Proforma:** This indicates that the financial statement is a projection of the future. These should be based on the best possible estimates at the time they are put together.

**Profit Margin:** This shows the percentage of gross income resulting in profits for the firm. Mathematically, find the value of net income plus interest expense minus the value of operator and unpaid operator labor and divide that value by gross income. Interest expense is added back to net income because it represents a return to the debt-financed assets. Removing the value of operator and unpaid operator labor shows that returns must be enough to cover this value.

**Rate of Return on Assets:** This shows the return to all assets employed in the business as a percentage of the total assets employed. Mathematically, it is found by dividing the numerator of net income plus interest expense minus the value of operator and unpaid operator labor by the denominator of average total agri-business assets. Interest expense is added back to net income because it represents a return to the debt-financed assets. Removing the value of operator and unpaid operator labor shows that returns must be enough to cover this value.

**Rate of Return on Equity:** This shows the returns to equity assets employed in the business as a percentage of the equity assets. Mathematically, it is found by dividing the numerator of net income minus the value of operator and unpaid operator labor by the denominator of average total equity assets. Removing the value of operator and unpaid operator labor shows that returns must be enough to cover this value.

**Evolving Line of Credit:** A type of credit account in which the borrower has a given credit limit which can be borrowed at any time. A credit card or standing account with an equipment dealer are examples of revolving credit lines.

**Risk Management**: Refers to any attempt to avoid the possibility of unfavorable outcomes under uncertainty. Insurance and buying on futures and options markets to lock in input or output prices are good examples of tools used in risk management.

**Sensitivity Analyses:** Refers to using alternative assumptions to determine what the outcome of a financial analysis will be if different outcomes are realized. For example, if developing a proforma income statement, using a range of assumptions associated with output prices helps to show how projected net income will change if prices differ from the base assumption.

**Standard Operating Procedures (Sops):** These are written sequences of steps required to perform a specific task. Milking a cow, for example, requires many steps. A written SOP allows the milker to perform this task in the same way every time the cow is milked.

**Statement of Cash Flows:** This shows cash income and cash expenses over a specified period of time, often a year. These receipts and payments are typically broken into three categories associated with operations, investments, and financing.

**Swot Analysis:** A systematic review of the firm's strengths, weaknesses, opportunities, and threats. This is used to draw focus on what the firm does well and what it may be able to do to take advantage of emerging market opportunities.

**Vision Statement:** Provides a summary of the firm's most important goals. Firms differ with respect to how specific they state these goals in their business plans. We recommend being as specific as you can comfortably be.

**Working Capital to Value of Production Ratio:** This represents working capital as a percentage of gross income. Working capital is equal to current assets minus current liabilities. Current assets include cash and other assets that will be converted to cash or used up within one year. Current liabilities are those that are payable within one year.

# Unit 6

# HRM in Agribusiness Expansion

## 6.1 Introduction to Agribusiness

Agri-business as a concept was born in Harvard University in 1957 with the publication of a book "A concept of Agri-business", written by John David and A. Gold Berg. It was introduced in Philippines in early 1966, when the University of the Philippines offered an Agri-business Management (ABM) programme at the under-graduate level. In 1969, the first Advanced Agribusiness Management seminar was held in Manila.

Agri-business is the sum total of all operations involved in the manufacture and distribution of agri-business supplies, production activities on the agri-business, storage, processing and distribution of agri-business commodities and items made from them.

The agribusiness approach is a method of examining agri-business problems in a new and more comprehensive setting. One benefit from this approach has been the release of workers, agri-business manpower from agriculture for employment in new non agri-business occupations including the armed forces during wars. This has resulted in tremendous economic growth and development and an improved standard of living.

Agribusiness consists of several million agri-business units and several thousand business units, each an independent entity, free to make its own decisions. Agribusiness is the sum total of hundreds of trade associations, commodity organizations, agri-business organizations, quasi-research bodies, conference bodies, and committees, each concentrating on its own interests. The U.S. government also is a part of agribusiness to the degree that it is involved in research, the regulation of food and fiber operations, and the ownership and trading of agri-business commodities. Land-grant colleges, with their teaching, experiment stations, and extension functions, form another sector of agribusiness. In summary, agribusiness exists in a vast mosaic of decentralized entities, functions, and operations relating to food and fiber.

The evolution from agriculture to agribusiness has brought with it numerous benefits. These include reduced drudgery for laborers; the release of workers for nonagricultural endeavors; a better quality of food and fibers; a greater variety of products; improved nutrition; and increased mobility of people. The release of agri-business manpower and the creation of new, off-the-agri-business jobs have been the basis for the country's economic growth and development for the last 150 years. The key to this growth and development has been increased worker productivity, which in turn spurs creativity, new products and wealth. This translates into risk capital, new factories, new jobs, and increased consumer purchasing power.

Agribusiness treats the different aspects of raising agricultural products as an integrated system. Agri-business owners raise animals and harvest fruits and vegetables with the help of sophisticated harvesting techniques, including the use of GPS to direct harvesting operations. Manufacturers develop more efficient machines that can drive themselves. Processing plants determine the best way to clean and package livestock for shipping. While each subset of the industry is unlikely to interact directly with the consumer, each is focused on operating efficiently in order to keep prices reasonable.

Market forces have a significant impact on the agribusiness sector. Changes in consumer taste alter what products are grown and raised. For example, a shift in consumer tastes away from red meat may cause demand (and prices) for beef to fall, while increased demand for produce may shift the mix of fruits and vegetables that agri-business people raise. Businesses unable to rapidly change in accordance to domestic demand may look to export their product abroad, but if that fails they may not be able to stay in business.

Countries with agri-business industries face consistent pressures from global competition. Products such as wheat, corn, and soybeans tend to be similar in different locations, making them commodities. Remaining competitive requires agribusinesses to operate more efficiently, which can require investments in new technologies, new ways of fertilizing and watering crops, and new ways of connecting to the global market. Global prices of agricultural products may change rapidly, making production planning a complicated activity. Agri-business people may also face a reduction in usable land as suburban and urban areas move into their areas.

## 6.2 Sectors

Input supplier and manufacturer

- ☆ Machinery dealer or fertilizer & chemicals supplier

- ☆ Fertilizer & chemicals manufacturer

Inputs are the various resources and services producers use to produce agricultural products. Using our definition, agribusiness firms include computer software developers, financial services companies, insurance providers, accountants and attorneys, as well as the more traditional agricultural input companies such as seed, feed, fertilizer, agri-business equipment, irrigation, animal pharmaceuticals, livestock handling equipment, and horticultural supplies.

Agri-business operation

- ☆ Cropping agri-business
- ☆ Livestock agri-business

First handler and processors

- ☆ Grain elevator
- ☆ Milk cooperative

Services provider

- ☆ Financial institution
- ☆ Consultant
- ☆ Integrated business

### Agri-business Involves Three Sectors

1. Input sector: It deals with the supply of inputs required by the agri-business for raising crops, livestock and other allied enterprises. These include seeds, fertilizers, chemicals, machinery and fuel.
2. Agri-business sector: It aims at producing crops, livestock and other products.
3. Product sector: It deals with various aspects like storage, processing and marketing the finished products so as to meet the dynamic needs of consumers.

Therefore, Agribusiness is sum total of all operations or activities involved in the business of production and marketing of agri-business supplies and agri-business products for achieving the targeted objectives.

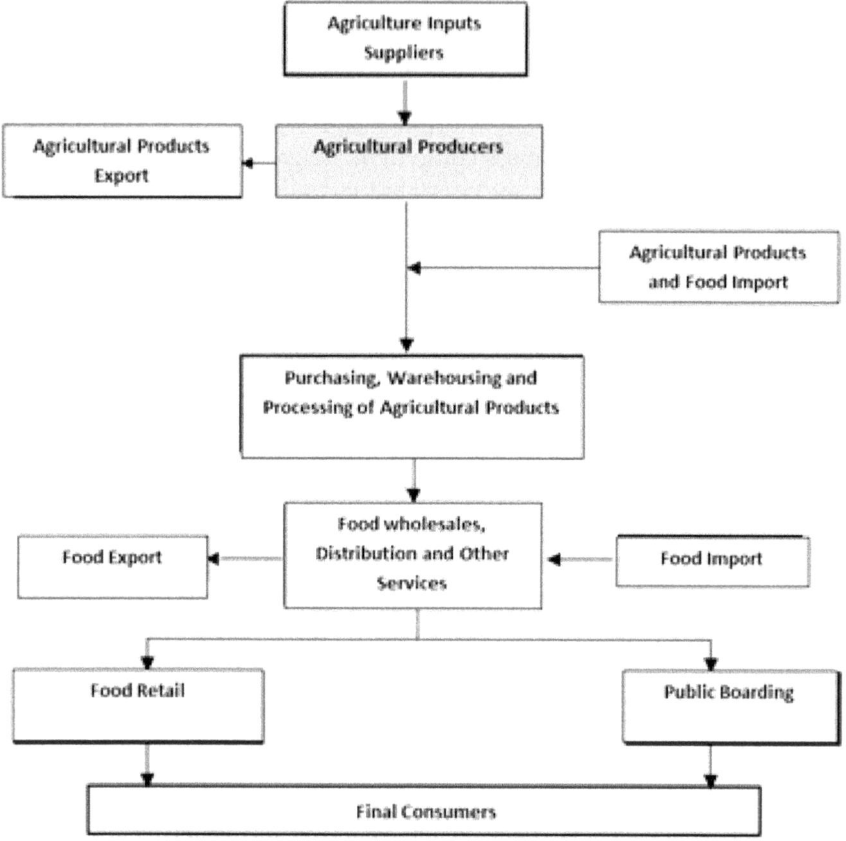


**Fig. 6.1: Basic structure of agribusiness**

## 6.3 Importance and Dimensions of Agri-business

1. It deals with agricultural sector and also with the portion of industrial sector, which is the major source of agri-business inputs like fertilizers, pesticides, machines, processing and post- harvest technologies.
2. It suggests and directs the government and private sectors for development of sub sectors.
3. It contributes a good part of the national economy

Agri-business includes all business enterprises that buy from or sell to agri-business owners. The transaction may involve either a product, a commodity or a service and encompasses items such as:

(a) productive resources e.g. feed, seed, fertilizers equipment, energy, machinery etc.

(b) agricultural commodities e.g. food and fibre etc.

(c) facilitative services e.g. credit, insurance marketing, storage, processing, transportation, packing, distribution etc.

Agri-business can also be defined as science and practice of activities with backward and forward linkage related to production, processing, marketing and trade, distribution of raw and processed food, feed, fiber including supply of inputs and service for these activities.

1. It deals with different components of both agricultural and industrial sector, their inter- dependence and influence of one sector on other.
2. It deals with decision making process of agri-business either private or government in relation to production and selling aspects.
3. It deals with strengths and weaknesses of a project and thereby their viability in competing enterprises. Agri-business is always market oriented. The Agribusiness sectors contribute over 60% of Sub-Saharan Africa economies and employ over 70 % of the population in most developing countries. This makes it necessary to understand the management of people in agribusiness sectors.

## 6.4 HRM in Agriculture

### 6.4.1 Introduction

There is increasing evidence and recognition that what matters for development, more than natural resources and man-made physical capital, is the capability of people to be effective and productive economic agents, in short, human capital. In the particular case of agriculture, most studies on the subject establish that the education and skills of agricultural people are significant factors in explaining the inter-agri-business and inter-country differences in agricultural performance, along with the more conventional factors such as availabilities of land and water resources, inputs, credit, etc.

With the shrinking of per caput agricultural resources following demographic growth, with the agricultural labour force in the developing countries projected to continue at positive (though declining) growth rates and with the share of young people in the total also continuing to grow, the task of upgrading the literacy, the skills and other capabilities of the agricultural people is enormous, for coping with both the increases in numbers and the backlog inherited from the past. Moreover, the increasingly binding character of natural resource scarcities imposes severe limits on the extent to which production increases can be had through expansion of extensive agriculture. The generation and diffusion of technology and management capabilities for more intensive and modernized agriculture and supporting services become imperative. This can only be achieved through the upgrading of the quality of human resources employed in agriculture.

It is noted that many dimensions of the human resources development issue are final end-objectives of development, e.g. Iiteracy, better health and nutrition, etc. Although this chapter is concerned with policies to upgrade the quality of people to become more productive and more energetic economic agents, the need to make progress in literacy, health, nutrition, etc., as objectives in their own right, should not be lost sight of. This is important, since it implies that evaluation of returns to investment in these areas must take into account the value of improvements in literacy, etc., as increasing the welfare of individuals directly and not only indirectly through making them more productive economically. These considerations cannot but influence the criteria for making decisions concerning the allocation of scarce resources, e.g. between promoting basic education versus creation of more directly productive agricultural skills.

### 6.4.2 Magnitude of the Task

A first impression can be had by observing that the present population economically active in agriculture in the developing countries of just over I billion is likely to continue to increase by some 13 percent in the next 20 years. The growth rate is slowing down from 1.2 percent p.a. in the last 20 years to 0.6 percent in the next two decades, and indeed the PEA is about to peak in the regions of Latin America/ Caribbean and East Asia. But this is not likely to happen in the two regions with the highest shares of their population in agriculture and with high incidence of rural poverty. This means that 20 years from now, these two sub-regions are still likely to have 60 percent of their labour force depending mainly on agriculture for employment and income. This contrasts with likely developments in the Latin America/Caribbean and Near East/North Africa regions which seem to be transiting towards patterns of labour force dependence on agriculture more typical of southern Europe.

Naturally, the human resource development effort has to provide for the entire agricultural, and indeed the rural, population, not only those classified as economically active. In particular, interventions in the areas of basic literacy, health and nutrition have to reach people well before they grow to become members of the PEA. The magnitude of the task can be appreciated from a few related parameters. In the first place, the numbers have to be multiplied by a factor of 2.2 for the developing countries as a whole to obtain the estimates of the total agricultural population. Secondly, the age structure of the rural population implies that some 13 percent of the total, or some 350 million, are in the age group 15-24 years, a group commonly referred as youth in the HRD programmes. Their numbers will be edging up towards 400 million in the future. Indeed, from the point of view of providing basic education services, these estimates will have to be more than doubled to account for children in the age agroup 6-15. Finally, the share of economically active women in the PEA is about 30 percent for the developing countries as a whole, but with wide regional variations, e.g. 56 percent in sub-Saharan Africa, 37 percent in Near East/North Africa, 31 percent in Asia, but only 12 percent in Latin America. It is obvious that the data referring to women are of great importance for focusing the HRD effort in the rural areas given the increasing recognition of the role of women-in-development in policy making in combination with the fact that past HRD policies have tended to favour men rather than women.

### 6.4.3 Basic Education and Agriculture

Basic education, often referred to as literacy and numeracy education, is the most fundamental of HRD efforts, not only as a universal right of the individual but also as the foundation for any further initiative in human resource development in agriculture designed to improve agricultural production and, hence, incomes and welfare. Basic education can improve significantly the efficacy of training and agricultural extension work which in turn affect agricultural production through: (a) enhancing the productivity of inputs, including that of labour; (b) reducing the costs of acquiring and using information about production technology that can increase productive efficiency; and (c) facilitating entrepreneurship and responses to changing market conditions and technological developments. The relationship between education and agricultural development cuts both ways and the two are mutually reinforcing, with demand for schooling rising as rural incomes increase.

### 6.4.4 Agricultural Extension and Training

Agricultural extension "assists agri-business people, through educational procedures, in improving agri-business methods and techniques, increasing production efficiency and income, bettering their levels of living and lifting the social and educational standards of rural life". Publicly supported agricultural extension services for agri-business people are an innovation of the twentieth century. For example, the USA established its Cooperative Extension Service in 1914.

This HRD innovation in agriculture has been spreading in recent years. Out of 198 extension organizations in 115 countries which provided reports to FAO in 1989, only 10 percent had been established before 1920, while 50 percent had been established after 1970. The increasing adoption of organized extension services in the developing countries reflects the realization of their importance for agricultural development and the high rates of economic returns that have been demonstrated by the experience of countries where extension has been properly delivered. For example, a study reports that in the USA a $1000 increment in extension spending was associated with a $2173 increase in agri-business output within a two-year period. Comparative studies from several countries provide further support to the finding of relatively high economic returns to investment in agricultural extension. More recently, studies on productivity-increasing effects of agricultural extension services have been reported. A report on World Bank support to agricultural extension services in 22 sub-Saharan African countries is illustrative. An average of 40 percent increase in yields in the first year has been recorded in a 1989-90 study.

The findings of the recent Global Consultation on Agricultural Extension of FAO indicate that the rate of return to investment in extension is influenced by a variety of factors. These range from the economic value of the agri-business product, with cash and export crop owners enjoying higher returns than food crop owners; the general economic climate, with returns being lower in relatively poorer agricultural nations; to the nature of the extension services provided, with larger returns to those systems that embrace large numbers of agri-business owners having lower costs. More generally, it is recognized that an effective extension system cannot be considered on its own and that it "needs a supportive environment that includes a long-term commitment to agricultural growth expressed through the provision of adequate agricultural support services-of which extension is but one- and macro-economic policies that, at a minimum, do not disfavour agriculture".

Agricultural extension services in the world have been expanding during the past three decades. Around 1959, there were approximately 68 organized extension services, with as many as 180000 agricultural extension personnel. By the year 1980, the number of organized extension services had increased to around 150, with a total personnel of about 350000. The estimated extension workers in 1989 stood at approximately 600000, of whom nearly two-thirds were located in developing countries. A least developed country like Mozambique, for example, had in 1989 about 350 professional/ technical staff in extension following the establishment of a national extension service in 1986 with UNDP/FAO assistance. The data collected for FAO's Global Consultation indicate that agricultural extension expenditure was approximately US$4.6 billion in the 98 countries for which data were available, of which nearly 87 percent was in developing countries. If it had been possible to include all countries of the world, the estimated total expenditure on extension would probably have exceeded US$6 billion per year.

In spite of the tremendous increase in the numbers of agricultural extension workers in the last three decades, the actual coverage of agricultural extension services in the developing countries has been limited. In the USA, Canada and Europe, one public extension agent covers about 400 economically active persons in agriculture, even before counting the services of private sector extension agents. In the developing countries an extension worker covers on average about 2500 such persons. This suggests that in actual practice, only one out of every five economically active persons in agriculture receives extension services in the developing countries. This rate is likely to be lower when one considers that about one-fourth of the extension worker's time is devoted to non-educational duties, which was equivalent to approximately 140000 full time years of extension workers' time in 1989.

Another issue is the kind of agri-business served by the extension agents. Data from the Report of the FAO Global Consultation on Agricultural Extension show that in the reporting developing countries, 6 percent of extension agents' time and resources is devoted to large commercial agri-business, 26 percent to smaller commercial agri-business, while 24 percent is devoted to subsistence agri-business and 6 percent to agri-business women. In a well-documented case study of the extension programme in two provinces of Turkey, however, the record indicates that 100 percent of the 5100 large-scale agri-business were served by the extension service, while only 55 percent of the 62 300 small-scale agri-business were receiving extension services. Of the 17900 medium-scale agri-business, 90 percent were receiving services from extension.

The preceding paragraph illustrates the problems associated with public extension services. The private sector has varying degrees of involvement in agricultural extension work in most countries. In developed countries, the trends towards privatization relate to budgetary problems. In the developing countries, the need to increase coverage and contain the costs of public extension are the main driving forces behind the increasing involvement of the non-governmental organizations (NGOs) and the private sector in extension. For example, in Colombia, 35 percent of the 2315 extension agents are provided by an NGO-the National Federation of Coffee Growers.

### 6.4.5 Technical and Professional Education in Agriculture

The amount and quality of trained technical and professional manpower in agriculture are critical factors, both in agricultural development and more general HRD. This "human capital" is relatively scarce because training takes years and is costly. However, investing in technical and professional education has a high multiplier effect when trained personnel are properly employed as extension agents, trainers, researchers, programme managers, policy makers and in the private sector.

Although many developing countries still have serious shortages of trained manpower in fields related to agriculture, considerable progress has been made during the last three decades. By 1983, for example, there were over 400000 trained agricultural personnel in 46 countries in Africa. In 25 of these countries, moreover, the existing institutional capacity was sufficient to allow the training of the required number of agricultural personnel for the year 2000. Worldwide, increases in institutional capacity for training are reflected in the increased number of extension personnel mentioned earlier, as well as of agricultural research personnel. ISNAR reports that agricultural research personnel in developing countries increased at the rate of 7.1 percent annually from 19753 to 77737 during 1961-65 to 1981-85.

However, when these numbers are compared with requirements, especially in the least developed countries, there are still considerable shortages. For example, Ethiopia would have to graduate 231 people at the professional level and 1254 at the technical level annually to reach the minimum estimated requirements for trained manpower in agriculture by the year 2000. As mentioned earlier, the problem in agricultural extension is a shortage of well trained extension agents in many developing countries. In the case of research, UNESCO data show that there were approximately 500 scientists and engineers per million population in a sample of developing countries as compared to more than 3000 scientists and engineers per million population in developed countries.

For most developing countries, the supply of technical and professional manpower in agriculture for the next decades will remain problematic. In Africa, 18 of 46 countries surveyed in 1983, had reported that their technical agricultural personnel was less than 50 percent of the year 2000 minimum requirement. Even in those developing countries where, in general, numbers of technical and professional manpower met the minimum requirement, the problem is an excess of numbers in certain fields and shortages in others. For example, in Africa only 7 percent of technical and professional trained manpower in agriculture are in forestry, 5 percent in fisheries and 11 percent in livestock.

The major problems of developing countries in the area of agricultural education and training as they face the new century include inadequate institutional capacity, relatively low level of public and private support to agricultural education, and limited resources and experience to cope with new areas of training in agriculture, *i.e.* environment and natural resource management, biotechnology, agri-business systems management and agribusiness.

## 6.5 Nature of HRM in Agri-business

The nature of the human resource management has been highlighted in its following features:

### 6.5.1. Inherent Part of Management

Human resource management is inherent in the process of management. This function is performed by all the managers throughout the organisation rather that by the personnel department only. If a manager is to get the best of his people, he must undertake the basic responsibility of selecting people who will work under him.

### 6.5.2. Pervasive Function

Human Resource Management is a pervasive function of management. It is performed by all managers at various levels in the organisation. It is not a responsibility that a manager can leave completely to someone else. However, he may secure advice and help in managing people from experts who have special competence in personnel management and industrial relations.

#### 1. Basic to all Functional Areas

Human Resource Management permeates all the functional area of management such as production management, financial management, and marketing management. That is every manager from top to bottom, working in any department has to perform the personnel functions.

#### 2. People Centered

Human Resource Management is people centered and is relevant in all types of organization. It is concerned with all categories of personnel from top to the bottom of the organization. The broad classification of personnel in an industrial enterprise may be as follows: (i) Blue-collar workers (*i.e.* those working on machines and engaged in loading, unloading etc.) and white-collar workers (*i.e.* clerical employees),(ii)Managerial and non- managerial personnel, (iii)Professionals(such as Chartered Accountant, Company Secretary, Lawyer, etc.) and non-professional personnel.

#### 3. Personnel Activities or Functions

Human Resource Management involves several functions concerned with the management of people at work. It includes man power planning, employment ,placement, training, appraisal and compensation of employees. For the performance of these activities efficiently, as separated apartment known as Personnel Department is created in most of the organizations.

#### 4. Continuous Process

Human Resource Management is not a one shot 'function. It must be performed continuously if the organizational objectives are to be achieved smoothly

## 6.6 Significance of HRM in Agri-business

Human resource management plays an important role in agriculture and food processing sector and represents one of the most complex issues in agro-food companies, being influenced more by social than economic determinants. The expansion of agricultural and food processing industry has been viewed as one of the most significant sources of economic growth in agro-food sector in developing countries, while the most significant contribution to the success of every agro-food organization has been the productivity of its employees. Today it is commonly accepted that HRM has a significant influence on overall success of agricultural and food processing companies.

The overall objective of every successful organization is to achieve sustained competitive advantage in a long-term. According to resource-based view, human resources may be a source of competitive advantage of a company, as companies may create economic value through utilization of their employees, only if their knowledge, skills and abilities can be used appropriately. Beside other resources, human resources are often seen as a crucial strategic resource used by companies to achieve this objective, especially in agricultural and food processing sector.

Having in mind that their people are their most valuable asset, agro-food companies and their managers need to search for the best ways to manage their employees effectively with an objective to achieve and maintain the desired market position. The role of human resource management in agro-food sector- where production, processing, and timely delivery of high-quality products depend on skills, knowledge and abilities of their employees- are of crucial importance. It is primarily aimed at recruiting, managing and maintaining the staff composed of highly specialized experts, semi-skilled workers and unskilled workers.

However, there is a lack of research regarding HRM in this sector as literature has paid insufficient attention to this topic. Research in HRM is usually focused on other sectors rather than agriculture and food processing. Additionally, it may happen that certain HRM practices developed for large companies cannot always be successfully implemented in smaller companies; also, practices defined for companies from other sectors may not work well in agricultural and food sector. Having this in mind agricultural (as well as food processing) managers often have little evidence to rely on when developing HRM policies and procedures for their companies. This makes HRM processes in these companies much more complex, as managers need to develop specific systems for managing the workers efficiently and effectively. This explains the importance of HRM role in agro-food organizations (particularly multinational ones, which face much more complex issues in managing their employees).

Having in mind that main challenges faced by agro-food companies include attracting, motivating, and retaining qualified employees, their HR departments should deal with different activities in order to satisfy and synchronize both individual and organizational goals. It should be ensured that each employee strives to achieve his/her goals which have derived from organizational objectives.

Since modern HRM concept was brought relatively recently, and most often by foreign-owned companies, it was assumed in this research that mainly the subsidiaries of these companies in Serbia have implemented modern HR processes, rather than local companies. This is the reason why this survey focused only on foreign-owned companies in agricultural and food processing sector (with many of them being subsidiaries of foreign multinational companies) in order to examine the following hypothesis:

***Hypothesis 1: Foreign-owned agricultural and food processing companies performing key HRM activities.***

Human resource planning refers to a process aimed at analysing and predicting organizational needs for human resources and their availability. Since success of companies involved in agricultural and food processing industry greatly depends on skills and experience of their human resources specialists, the objective of this process is to identify the number and the types of employees needed by an organization in the future to fill in its vacancies, as well as changes the organization should undertake in order to accomplish its goals – in terms of reducing the number of employees, training existing employees or hiring new personnel.

In order to achieve expansion in agro-food organizations, HR departments need to perform appropriate personnel planning in all departments. HR department should work together with line managers in order to predict future organizational HR needs (both short-term and long-term) and determine the number and qualifications of specialized workers the organization will require depending on identified needs. It is the duty of HR department to ensure that these requirements are met in order to support crucial agricultural or food processing processes in the company.

When specific HR plans have been made a company may start with recruitment- the process in which organization looks for candidates for potential employment. Therefore, this process refers to identification and attracting potential job candidates in order to satisfy current and future organizational HR needs. The aim is to obtain the necessary number and quality of employees in order to accomplish organizational HR needs. A key challenge agricultural and food processing companies must face in attempt to become and remain competitive refers to deciding on the ways of attracting and retaining skilled workforce.

Companies in agrofood sector employ different types of workers- mainly seasonal and contract workers, technicians, but also supervisors and managers and thus require a wide dispersion of skills levels- from unskilled and low-skilled to high-skilled jobs. As observed, skills such as creativity and critical thinking today become more important in agricultural sector. Having this in mind, skills and labour shortages may be a serious issue in this sector causing difficulties in recruiting adequate candidates. Besides, one of main problems faced by agricultural companies refers to ageing workforce within agriculture as well as their education level.

Another challenge which may be faced refers to too many job candidates in agricultural sector, having in mind that the only option for most inhabitants in rural areas is to work in this sector. Having in mind the specific nature of the business as well as knowledge, skills and abilities required, HR department should cooperate with department managers in the process of recruitment of agricultural and food processing employees.

Companies provide training programs for their employees to enable them to gain specific knowledge and skills necessary for performing their jobs effectively. HR department should pay significant attention to these issues. Whenever it is noticed that an employee lacks specific skills or knowledge training programs should be arranged – within the company or in other organizations, universities or institutes; employees may be sent to different seminars, conferences or workshops in order to gain necessary knowledge and become familiar with the best practices which they may later implement properly in the organization. In agriculture and food processing companies training represents one of the most significant responsibilities at all organizational levels. Employees need to be trained to use agricultural machines or perform risky processes in food production or processing, particularly when a new technology has been introduced. Safety trainings represent a priority for human resource management, in order to eliminate or reduce the possibility for injuries in the workplace or even death caused by inadequate handling with equipment and machines, or work-related illnesses.

HRM activities may be performed within the company or outsourced. There are many reasons for outsourcing these activities: it may be too complex for companies to deal with all HRM activities themselves; companies may reduce costs of HRM (of employing highly trained and experienced HR employees as well as necessary office space); they may improve the quality of HRM processes (services provided by HR agencies are professional as they are specialized only for particular HRM processes). Outsourcing may be a good solution for companies which don't have enough experience in HRM

However, from a long-term perspective companies often prefer to establish their own HRM departments, rather than engage HR agencies for performing their crucial HRM activities.

The literature has identified various factors which are expected to have an influence on the decision to perform HRM processes within the company or outsource. Thus the following hypotheses were defined and later examined for five different HRM activates: HR planning, recruitment, selection, and training as well as performance appraisal of employees.

***Hypothesis 2: The decision on performing key HRM activities themselves or outsourcing them depends on country of origin of the company.***

***Hypothesis 3: The decision on performing key HRM activities themselves or outsourcing them depends on company size.***

***Hypothesis 4: The decision on performing key HRM activities themselves or outsourcing them depends on company age.***

***Hypothesis 5: The decision on performing key HRM activities themselves or outsourcing them depends on the size of HR department.***

## 6.7 HR in Agribusiness Expansion

The role of human resources in agribusiness expansion is focused on recruiting and managing a staff composed of both highly specialized professionals, semi-skilled laborers and unskilled labors.

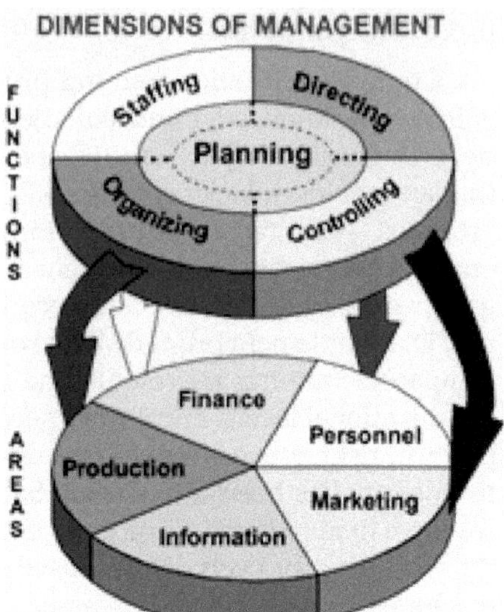

Fig.6.1: Dimensions

Agribusiness includes technical and labor-intensive activities that are required to optimize production from the inputs that are invested in crop production and animal husbandry. Agribusiness expansion, therefore, requires HR departments to conduct proper planning of workforce requirements in all sectors of crop and livestock production.

### 6.7.1 Forecasting Human Resources Needs of the Agribusiness

The HR office works closely with the operations management team to determine the adequate number of specialized workforce and agri-business support laborers that will be required to handle the increased volumes of agri-business activities and produce marketing. To this end, HR plays the role of ensuring that the personnel requirements for supporting the different technical agricultural processes and marketing activities of the agribusiness are adjusted to match the additional activities that come with expansion. HR either multi-tasks the workforce or recruits specialized employees to ensure that the skills, agri-business machinery and labor-intense activities are supplied with adequate number of personnel.

### 6.7.2 Retaining and Recruiting Personnel

HR collaborates with the heads of sections or departments to maintain and recruit personnel but also to acquire the desired skill sets for an agribusiness workforce. HR must collaborate with the department heads because of the varied nature of the business and the skills required of the workforce. Managers of areas such as horticulture, livestock breeding, milk production, milk processing and marketing are in the best position to determine the minimum requirements and suitable remuneration packages for the specialized agricultural workers of their respective departments.

### 6.7.3 Implementing Policies

HR also plays the role of ensuring that the expansion of the agribusiness is done within the established policies for compensating, promoting and motivating employees in different agri-business departments and production sections. HR ensures that the job descriptions and responsibilities of each employee are aligned to the overall production and performance targets of the business. HR collaborates with top management to ensure that employees participate fully and benefit in the growth activities, such as research and development of animal breeds and crop species.

### 6.7.4 Managing Specialized Agribusiness Knowledge and Skills

Managing knowledge involves facilitating employees to enhance their agricultural skills and work capabilities through employee mentoring programs, research and development in new animal breeds and crop species and education advancement opportunities in agricultural fields of specialization such as agricultural engineering. The dynamic nature of agribusiness requires a personnel team capable of adjusting and living up to expectations of the ever-changing business environment. HR, therefore, spearheads the efforts of supporting and keeping track of the knowledge management initiatives to ensure that personnel are continuously exposed to learning opportunities on work processes and agricultural technological developments.

## 6.8 Writing Agribusiness Plan

A business plan is a description of a business in terms of its goals, reasons for existence and strategies for attaining the objectives.

### 6.8.1 Components of the Plan

#### 1. Business Background and Objectives of the Business

This section provides background information about the agribusiness. It should include:

- ☆ Status and form of the agribusiness (old or new and the legal description; sole proprietorship, partnership, cooperative or company).
- ☆ Physical location of the agribusiness.
- ☆ Vision, mission, goals and philosophy of the agribusiness.
- ☆ Opportunity and viability of the agribusiness.

#### 2. Products and Services Offered by the Business

This section describes the products and or services to be offered by the agribusiness. The factors which will give one a competitive advantage or disadvantage, for example, the level of quality or uniqueness or proprietary features. What are the leasing structures of the products or services if any?

#### 3. Marketing Strategy and Plan

- ☆ This section describes the market share that the agribusiness expects to control at the planning period and how much the share should grow annually over the plan period.
- ☆ It further indicates the agribusiness' ability to penetrate the market as planned in terms of sales revenue, profits, growth rate and trends.
- ☆ What share of the market will one have? What is the current demand in the target market? These are growth trends, trends in consumer preferences

and trends in product development, growth potential and opportunity for the business.

***Marketing Plan***

This entails:

- ☆ How will the product be priced relative to the competition, including customer credit policies?
- ☆ How products will reach target customers and coordination mechanism of the actors in a distribution channel.
- ☆ What communication methods will be used to ensure that target consumers of agribusiness products are continuously informed in a positive manner? This should include the promotional budget for that purpose.
- ☆ People and processes applicable for agribusiness services. What processes will be put in place for service delivery and how will people be involved in that process?

### 4. Operational Plan

This section explains how the agri-enterprise will produce its products. It should include physical location, facilities, equipment, a description of the production process and quality control.

### 5. Financial Plan

The projections need to be for a period of three to five years. The items covered in this section are:

- ☆ Balance sheet (statement of the financial position as at a particular date).
- ☆ Income statement (profit and loss account).

### 6. Agribusiness Management and Organisational Structure

This section outlines who will manage the agribusiness on a day-to-day basis. In case the employees are to be out-sourced, indicate the recruitment procedures.

### 7. Monitoring, Evaluation and Controls

All the activities of the agribusiness must be continuously monitored and regularly evaluated to ensure deviations from the plan are rectified promptly.

### 8. Risks and Assumptions

This section identifies the potential risks to the agribusiness. It shows your preparedness in dealing with unforeseen events and ways to mitigate them, including any underlying assumptions.

### 9. Other Details

Brochure and advertising materials, industry studies, blueprints, plans and production machinery. Photos of location and any other materials needed to support the assumptions in the plan.

# Questions

## Short Answers

1. What is agri-business?
2. what are HR policies in agri-business?
3. Explain Human Resource Strategy?

## Long Answers

1. What are the different sectors in agri-business?
2. Explain about basic structure of agribusiness?
3. Why Human Resource Management is known as pervasive function of management?
4. Explain the significance of HRM in agri-business

**Unit 7**

# Agribusiness Financial Management

## 7.1 Agribusiness Financial Records

### 7.1.1 Record Keeping

If a group of agri-businessman were to asked what task they dislike the most, many of them would probably say: "Doing the bookkeeping!" Unfortunately, you will not find any advice here that will make this chore fun, but it is absolutely essential in ensuring your success in business. Many small business people keep only enough information to satisfy themselves. This practice completely ignores the value of good, timely financial and production records in agri-business management. How will you know that feed expenses are rising higher than they should be? How will you know if you can afford to buy that new tractor? How will you know if you are making money?

Using a record keeping system designed for your business will help you to identify problems and solve them before they become unmanageable. These systems do not have to be complicated or expensive – in fact, a notebook with different pages for each of your income and expense accounts, may be all you need. More likely, however, you will want to have a system that also keeps track of your production assets, liabilities and ownership investment. There are many bookkeeping/ accounting software packages on the market that make this task easier and allow you to receive the management reports needed to make good decisions. Do not hesitate to ask for help from an accountant or bookkeeper, especially as you set up your system. They can give you advice that can save you time and money in the long run. Do not fret. You should consider this an investment in the business, which is every bit as important as that new cow shed or tractor.

### 7.1.2 Budgeting

Agri-business is a business with plenty of uncertainties and unknowns. How will the weather affect the crops? What will the interest rates be next year? Will the prices of my product be higher or lower than last year? Despite these questions and regardless of the size of your agri-business, budgeting is an important tool in successfully managing the business.

Most businesses have a five-year budget to plan long-term expenditure on equipment, land and buildings. The first year is very detailed and is used to control expenditure and identify financing requirements. But you have more important questions to answer before getting into agri-business. Can you afford to buy or start a agri-business? How much money will you need to borrow for the agri-business and the first few years of operation? What size agri-business do you need to provide the income you wish to have?

Depending on the type of agri-business business you plan to have, the projections should be five or more years into the future. For example, an apple agri-business starting from scratch requires statements for eight to 10 years since it takes four to five years before a new orchard starts producing.

### 7.1.3 Balance Sheet

A Balance Sheet is a summary of what your business owns (assets) and how those items were financed. It is 'balanced' because assets = liabilities + equity The first balance sheet you prepare should represent your financial state on the first day of business. At the end of each projected year of business, another balance sheet should be prepared based on the cash flow and income statements of each of those periods.

### 7.1.4 Cash Flow Statement

The cash flow statement is simply a projection of the timing and amount of cash flowing into the business from all sources, including loans, sales, government grants and owner's contributions, and the timing and amount of cash flowing out of the business in operating expenses, capital purchases, loan payments and withdrawals for living costs. These statements are prepared for each year divided into weekly, monthly or quarterly periods. Many agri-business will project cash flow out to five years. These projections are necessary to determine if sufficient cash will be generated and available from the business to cover expenditure. The time and amount of cash deficits can thus be measured and you can then negotiate an operating line of credit at the bank or other lending institution well ahead of time. The cash flow also provides information on cash balance, operating loan and equity adjustments for the balance sheets.

### 7.1.5 Income and Expense Statement

The final statement needed to complete the trio of financial records used to project the business into the future is the Income and Expense Statement (more commonly called the Income Statement). It provides an estimate of the net income for each year of the business. Although many of the items in this statement are included in the cash flow, it is important to recognize that the calculation of net income includes some non-cash items, such as depreciation on buildings and equipment, and excludes some cash items, such as principal payments on loans, capital purchases and sales and owner's withdrawals.

### 7.1.6 Financial Analysis

Simply preparing the series of statements is not enough. You need to analyse the results to ensure that you are within reasonable parameters in progress towards your goals. Once you are in business these same techniques are used to analyse past performance, too. The three most important areas for analysis are liquidity, solvency and profitability.

## 7.2 Agribusiness Financial Records

Agri-business managers use records to construct balance sheets, cash flow and income statements, and other financial aids for making more informed decisions in such areas as machinery purchases, adding or deleting enterprises, size expansion, etc. Some lending agencies and governmental bodies require financial and/or production records be maintained over a number of years. For example, the government agri-business program requires certain production and acreage records be reported and maintained by the agri-business owner. Also, "planning" for conservation compliance and other aspects of soil and water management essentially become historical records over time.

Record-keeping refers to keeping, filing, categorizing and maintaining agri-business financial and production information. Record-keeping can be accomplished through a variety of methods, from a basic hand record-keeping method to an elaborate computerized system.

Record analysis refers to evaluating agri-business records. The evaluation process allows a agri-business manager to make informed decisions based on actual agri-business performance. Obviously, record analysis cannot take place without first keeping records. Therefore, establishing and using an effective agri-business record-keeping system for an ongoing agri-business operation aids in agri-business planning, informed decision-making and analysis of both production and financial records.

On agri-business, there are two distinct types of records—financial and production. Financial records relate primarily to money or the financial interactions of the agri-business. Financial records justify or prove agri-business income or expense transactions. Product sales, operating expenses, equipment purchases, accounts payable, accounts receivable, inventories, depreciation records, loan balances and price information are all examples of financial records.

Production records are items that relate to quantities of inputs and levels of production by enterprise and/or by resource type. They consist of crop yields, plant populations, calves born, pounds of milk produced, weaning weights, death loss, etc.

Both production and financial records are important to the efficient management of today's agri-business business. When such information is accurately maintained and categorized, it can be used to produce useful decision-making information.

**Fig. 7.1: Record keeping**

## 7.2.1 Selecting a Record-Keeping System

Selecting a record-keeping system should depend on the expected use of the records. There is no "best" record keeping system for all situations, but, at minimum, a agri-business records system should:

- ☆ Provide accurate and necessary information
- ☆ Fit into the agri-business organization or framework
- ☆ Be available in a form to aid decision-making

The person responsible for keeping the records should develop a habit of regularly and accurately posting transactions. Making all financial transactions through a bank (checking) account can be useful. For an accuracy check, the monthly statement should be reconciled with the checkbook and record-keeping system.

A double-entry accounting system provides the most detailed accounting of agri-business business transactions. A significant amount of time is usually needed to learn and implement such a system. The simpler cash accounting system, with inventory adjustments, will suffice for most agri-business operations, and is an accepted method of reporting income and expenses for tax purposes.

### Comparing the Hand and Computer System

The use of computers and computer software has expanded on agri-business in recent years. However, a hand recording system is still useful for many agri-business owners. When selecting a record-keeping system, both hand and computer systems should be considered. Some characteristics of each are as follows:

**Fig. 7.2: Manual calculating the records data**

- ☆ Low initial out-of-pocket expense
- ☆ Easy to implement
- ☆ Time-consuming
- ☆ More opportunities to make mistakes

- Limited in extent of analysis without extraordinary investment of time and effort
- Higher initial out-of-pocket expense
- May require significant amount of study
- Fast
- Accurate
- Can be a powerful analysis tool

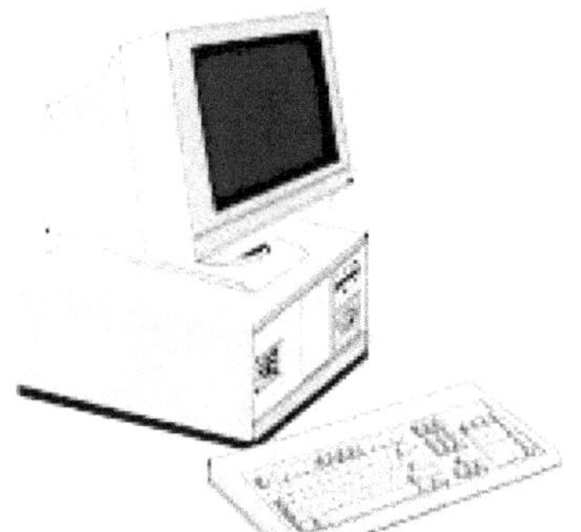

**Fig. 7.3: Computerised calculating system**

**Table 1. Example of Agri-business Record Keeping**

| | | Income | | | Expense | | |
|---|---|---|---|---|---|---|---|
| **Date** | **Description** | **Calves** | **Cull Cows** | **Corn** | **Feed** | **Supplies** | **Fuel** |
| 10/1 | Sold 10 calves @ 500 lbs/hd | $4,000 | | | | | |
| 10/1 | Purchased feed (2 tons) | | | | $475 | | |
| 10/3 | Sold 2 cows @ 1050 lbs/hd | | $1,250 | | | | |
| 10/6 | Purchased supplies | | | | | $165 | |
| 10/8 | Purchases 1500 gals of diesel | | | | | | $1,400 |
| 10/11 | Sold 2000 bu of corn | | | $4,600 | | | |

A large number of hand systems are available. One of the simplest systems involves the recording by hand of all financial transactions in a journal format. Purchase and sales activities are listed by hand as they occur. The entries show 1) the date, 2) the item involved (quantity, size, etc.) and 3) cash involved in sale or purchase. For example, table 1 portrays a agri-business record keeping system. The date and a short description of each transaction are listed first, followed by the dollar amount of each transaction under the appropriate income or expense category. The number of income and expense categories depends on the amount of specificity desired by the agri-business manager.

Table 2 is similar to table 1, but adds an enterprise accounting section. In addition to the income and expense categories, the transaction is also listed in an enterprise category. For example, the "purchased supplies" expense of $165 is first listed under the "supplies" category. But $125 of the expense is listed under "Cow/calf" expenses, and $40 is listed under "Corn" expenses. Again the number of enterprises used depends on the amount of detail desired by the agri-business manager.

**Fig. 7.4: Record Out System**

**Figure 2. Example of Enterprise Record Keeping**

| | | Income | | | Expense | | | Enterprises | | | |
|---|---|---|---|---|---|---|---|---|---|---|---|
| | | | | | | | | Cow/calf | | Corn | |
| Date | Description | Calves | Cull Cows | Corn | Feed | Supplies | Fuel | Income | Expense | Income | Expense |
| 10/1 | Sold 10 calves @ 500 lbs/hd | $4,000 | | | | | | $4,000 | | | |
| 10/1 | Purchased feed (2 tons) | | | | $475 | | | | $475 | | |
| 10/3 | Sold 2 cows @ 1050 lbs/hd | | $1,250 | | | | | $1,250 | | | |
| 10/6 | Purchased supplies | | | | | $165 | | | $125 | | $40 |
| 10/8 | Purchased 1500 gals of diesel | | | | | | $1,400 | | $200 | | $1,200 |
| 10/11 | Sold 2000 bu of corn | | | $4,600 | | | | | | $4,600 | |

To learn how the agri-business business stands on a cash basis, income and expense categories can be totaled weekly, monthly or during any time period desired. The difference in the two totals is the cash balance.

A number of record-keeping manuals or record books can be obtained from private or government lending agencies. Regardless of the system selected, entries should be made regularly. In addition, disciplining oneself to make every transaction through a bank checking account will ease record-keeping difficulties. Reconciling bank statements with agri-business records insures accuracy.

## Computerized Record-Keeping System

Historically, many agri-business managers have found keeping and analyzing financial records a challenge. However, a number of challenges have been addressed through computerized record-keeping systems. The advantages of using such a system depend on the expectations of the accounting system, the amount of time available to keep records and the attitude toward initial investment costs.

**Fig. 7.5: Using computerized accounting**

While a computerized record keeping system may make entering income and expense items easier, it may also take several days to learn the computer software. There is also a cost to purchase both the computer and the record-keeping software.

In a 1994 report, Pena, *et. al.*, evaluated six computerized agri-business record-keeping programs. The range of prices among the six programs was $6.95 to $79.95. Each program had unique features, but all six performed the basic recordkeeping functions.

Selecting a computerized record-keeping program should be done on the basis of features needed. Some programs will allow for enterprise accounting. Some programs will calculate payroll reports for employees. Many will have income tax options, but may vary on the completeness of the tax information supplied. Few financial record-keeping programs allow for production records to be kept simultaneously with financial records. For example, in many programs, sales of grain or livestock can be reported in dollars only, with no accounting for bushels or pounds.

A computerized record-keeping system will not necessarily save time. Its real advantage is in record analysis. Once the information is posted in the computer software, reports and analyses can be created, changed and printed. Computerized systems quickly and accurately sort and report a great deal of information. They can also provide monthly or annual summaries for identifying strengths and weaknesses of an operation.

If a hand system can provide the detailed information required by the manager to make quality agri-business decisions, it may be the best choice. However, if a hand system does not give the desired level of financial information, a computerized system should be considered.

## Analyzing Agri-business Records

Once agri-business record-keeping system has been established, analyzing the records can begin. Decision-making can be greatly enhanced by analyzing both production and financial records and their impact on profitability.

A number of financial analysis tools can be used when accurate and complete agri-business records are available. These tools include the balance sheet, income statement and projected monthly cash flow statement (including family living expenses). These three financial statements provide information for making short and long term financial decisions.

**Table 3. Financial Statements and List of Required Records**

| Statement | Records Required |
|---|---|
| Balance Sheet | Agri-business Assests Cost & Value |
| | Agri-business & Personal Asset Changes |
| | Livestock, Crop & Other Product Inventories |
| | Loan Balances |
| Cash Flow | Agri-business Income and Expenses |
| | Non-Agri-business Income & Expenses |
| | Debt Payments |
| Enterprise Analysis | Agri-business Income and Expenses |
| | Livestock & Crop Yields |
| Income Statement | Agri-business Income & Expenses |
| | Interest Payments |
| | Livestock & Grain Inventories |
| | Accounts Payable & Receivable |
| Income Taxes | Agri-business Income & Expenses |
| | Non-Agri-business Income & Expenses |
| | Interest Payments |
| | Depreciation Schedule |

The balance sheet gives the agri-business manager a "snapshot" of the net worth on a specific date. The net worth is the value of all assets on the agri-business less the amount of money owed against those assets.

Year-to-year profits are calculated on the income statement, also known as the profit/loss statement. The income statement is used to calculate net cash income, adjusted by changes in inventories and capital items.

The projected monthly cash flow statement is used to look ahead to the next year of operations. By projecting a cash flow for the next year, potential cash shortfalls can be noted and appropriate changes in the agri-business operation can be analyzed.

Table 3 was constructed to illustrate the kinds of records that go into the making of financial statements and production summaries. The left column contains the financial or production statement desired by the agri-business manager. The right column contains the records required to complete the statement. Records can be used for more than one statement. For example, "Debt Payments" are used in the Income Statement, Cash Flow Statement and in Income Taxes. Likewise, "Agri-business Income and Expenses" are used in the Income Statement, Cash Flow Statement, Enterprise Analysis and Income Taxes.

## 7.3 Types of Agribusiness Financial Statements

In order to help you to understand and manage money from your agri-business practice, your financial records can help you in securing loans, attract investment partners and communicate essential financial information to your business partners. The three most common types of financial statements required in agri-business are cash flow statements, income statements and balance sheets.

### 7.3.1 Cash Flow Statements

Cash flow statements show how much cash your agri-business has on hand. Small businesses can benefit from predicting their cash flow as well as just documenting it. A cash-flow projection is like a month-by-month budget that enables you to look ahead and plan for anticipated income and expenses. For instance, doing work for which you send a claim to an insurance company does not generate immediate income, but you can project when you will receive that payment.

Likewise, charging a piece of equipment to your company credit card does not take money out of your checking account immediately, but you know when to anticipate having to pay the expense. These types of anticipated income and expense are called accruals.

### 7.3.2 Income Statements

Income statements show your agri-business revenues, expenses and income, resulting in a profit or a loss. Income statements are particularly important if you need to borrow money because they tell lenders whether your business is making money or not. They also come in handy when you're calculating your quarterly estimated taxes and preparing your Schedule C for your federal income tax return. An income statement shows the following categories of financial information:

- ☆ Revenues: Examples include income received from massages, product sales, as well as from associates who share space with your practice.
- ☆ Expenses: Examples include fees paid to a bookkeeper, rent, purchase of supplies, utilities, memberships and training.
- ☆ Net Profit (or Loss) after Tax: Total revenues minus total expenses equals your profit or loss.

### 7.3.3 Balance Sheets

Balance sheets show your agri-business practice's assets, liabilities and equity. A balance sheet includes capital assets, such as equipment you own and security deposits that will be refunded to you later. You should update your balance sheet quarterly or annually. The balance sheet includes three categories of information:

- Your business assets: Examples include cash, furniture, equipment, supplies—essentially, everything you need to operate your agri-business.
- Your business liabilities: Examples include loan payments, utility bills, your office lease—essentially, everything your agri-business owes.
- Your business equity: Equity is the value that remains after you subtract the total liabilities from the total assets. The concept of equity reflects an accumulation of the owner's investments, withdrawals, and the accumulated earnings of agri-business.

## 7.4 Financial Management

### 7.4.1 Meaning and Definition

Finance is an integral part of the overall management rather than the fund raising activities. It is connected with all financial activities of planning, raising, allocating and controlling. The scope of financial management is very wide. Financial management holds the key to all activities. Finance may be regarded as a science, for it is a systematized body of knowledge of the phenomenon of the payment system. Finance is the management of monetary affairs of a company. It includes determining what has to be paid for and when, raising the money on the best terms available and diverting the available funds to the best uses.

Financial Management is an integral part of general management. It concerns managerial decision making. It helps in allocation of future financial requirements, allocation of resources and appraisal of financial problems.

- Financial management deals with how the corporations obtain the funds and how it uses them.
- Financial Management is the application of planning and control functions to the finance function.
- Financial management may be considered to be the management of the finance function.
- Financial management is the area of business management devoted to a judicious use of capital and a careful selection of sources of capital in order to enable a business firm to move in the direction of reaching its goals.

**Scope of Financial Management**

- Forecasting [estimation of the financial requirements]
- Financing [acquisition of capital]
- Allocation of funds
- Investment of funds
- Raising funds
- Co-ordination and control of funds [capital budgeting]
- Profit planning and control

- Decision Making
- Financial decisions
- Investment decisions
- Working capital decisions
- Dividend decisions

The important management functions are production, marketing and other functions. There exists inseparable relationship between finance and the other functions. Almost all business activities, directly or indirectly involve the acquisition and use of funds.

The finance function of raising and using money although has a significant effect on the other functions, yet it need not necessarily limit or constrain the general running of the business.

The functions of raising funds, investing them in assets and distributing returns earned from assets to shareholders are respectively known as financing decision, investment decision and dividend decision. A firm attempts to balance cash inflows and outflows while performing these functions. This is called liquidity decision.

The finance function includes:

- Long term asset-mix or investment decision
- Capital mix or financing decision
- Profit allocation or dividend decision
- Short term asset mix or liquidity decision

Finance functions call for skilful planning, control and execution of a firm's activities.

### *1. Investment decision*

- A firm's investment decision involves capital expenditure.
- A capital budgeting decision involves the decision of allocation of capital or commitment of funds to long term assets that would yield benefits (cash flows) in the future.
- Important aspects of investment decisions:
- The evaluation of the prospective profitability of new investments.
- The measurement of a cut off rate against the prospective return of new investments could be prepared.
- Investment proposals should be evaluated in terms of both expected return and risk.
- It also includes replacement decision *i.e.* decision of recommitting funds when an asset becomes less productive or non-profitable.
- The opportunity cost of capital is the expected rate of return that an investor could earn by investing money in financial assets of equivalent risk.

### 2. Financing Decision

Financial manager must decide when, where, from whom and how to acquire funds to meet the firm's investment needs. To determine the appropriate proportion of equity and debt is the main aim. The mix of debt and equity is known as the firm's capital structure. The finance manager must strive to obtain the best financing mix or the optimum capital structure for his firm. Capital structure is considered optimum when the market value of shares is maximized.

The change in the shareholder's returns caused by the change in profits is called financial leverage.

There must be a proper balance between return and risk. When the shareholder's return is maximized with given risk, the market value per share will be maximized and the firm's capital structure will be considered optimum.

In practice, a firm considers many other factors such as control, flexibility, loan covenants, legal aspects etc. in deciding the capital structure.

### 3. Liquidity Decision

Investments in current assets affect the firm's profitability and liquidity. Current assets should be managed efficiently for safeguarding the firm against the risk of liquidity. Lack of liquidity in extreme situations can lead to firm's insolvency.

A conflict exists between profitability and liquidity while managing current assets. If the firm doesn't invest sufficient funds in current assets, it may become illiquid and risky and would lose profitability, as idle current assets would not earn anything.

The profitability-liquidity trade off requires that the financial manager should develop sound techniques for managing current assets. He should estimate firm's needs for current assets and make sure that funds would be made available when needed.

### 4. Dividend Decision

- ☆ The financial manager must decide whether the firm should distribute all profits or retain them, or distribute a portion and retain the balance. The proportion of profits distributed as dividends is called the dividend-pay off ratio.
- ☆ The dividend policy should be determined in terms of its impact on the shareholder's value.
- ☆ The optimum dividend policy is one that maximizes the market value of the firm's shares.
- ☆ Dividends are generally paid in cash. But a firm may issue bonus shares.
- ☆ The function of financial management is to review and control decisions to commit or recommit funds to new or ongoing uses. Thus, in addition to raising funds financial management is directly concerned with production, marketing and other functions within an enterprise whenever decisions are made about the acquisition or distribution of assets.

## 7.4.2 Objectives of Financial Management

Financial management evaluates how funds are used and procured. The core of financial policy is to maximize earnings in the long run and optimize them in the short run. Financial management is an improved resource, mainly capital funds. The firm's investment and financing decisions are unavoidable and continuous. In order to make them rationally, the firm must have a goal. A firm's financial management may have the following as their objective:

- Maximization of firm's profit
- Maximization of firm's wealth

### 1. Maximization of Firm's Profit/ Profit Maximization

The maximization of profit is often considered as an implied objective of a firm. To achieve the aforesaid objective various types of financial decisions may be taken. Firms producing goods and services may function in a market economy. In a market economy prices of goods and services are determined in competitive markets. Firms in the market economy are expected to produce goods and services desired by society as efficiently as possible.

- Price system is the most important aspect of market economy which indicates what goods and services society wants.
- Higher demand for goods and services leads to higher prices resulting in higher profit for firms. It attracts other producers due to which competition in the market increases. An equilibrium price is reached when the supply of goods in a market matches the demand for those goods. The prices and profits of those goods and services tend to fall which has no demand by the society. Prices are determined by the demand and supply conditions as well as the competitive forces and they guide the allocation of resources for various productive activities.
- Profit maximization implies that a firm either produces maximum output for a given amount of input or uses minimum input for producing a given output. It is assumed that profit maximization causes the efficient allocation of resources under competitive market conditions and profit is considered as the most appropriate measure of a firm's performance.

### 2. Objections/ Criticism to Profit Maximization

This objective has been criticized. It is argued that profit maximization assumes perfect competition and in the phase of imperfect competition it falls to achieve its goal.

In the new business environment, profit maximization is regarded as unrealistic, difficult, inappropriate and immoral. There is a possibility of production of goods and services that are wasteful and unnecessary from the society's point of view. Also, it might lead to inequality of income and wealth. Firms producing same goods and services differ substantially in terms of technology, costs and capital. In such conditions, it is difficult to have a truly competitive price system and thus, it is doubtful if the profit maximizing behaviour will lead to optimum social welfare.

### 3. Wealth Maximization

Wealth maximization objective is as important as profit maximization. The operating objective of financial management is to maximize wealth or NPV (Net Present Value) of a firm.

The wealth of owners of a corporation is maximized by raising the price of the common stock. This is achieved when the management of a firm operates efficiently and makes optimal decisions in areas of capital investment, financing, dividend and current assets management.

The market price of a firm's stock represents the focal judgment of all market participants as to what the value of a particular firm is. It takes into account present and prospective future earnings per share, the timing and risk of these earnings, the dividend policy of the firm and many other factors that bear upon the market price of the stock.

The value/wealth maximization objective of a firm is superior to profit maximization objective due to the following reasons:

- ☆ The value maximization objective of a firm considers all future cash flows, dividends, EPS, risk of a decision etc. whereas profit maximization objective does not consider the effect of EPS, dividend paid or any other returns to shareholders.
- ☆ A firm that wishes to maximize the shareholder's wealth may pay regular dividends, whereas a firm that wishes to maximize profit may refrain from paying dividend payment to its shareholders.
- ☆ Shareholders would prefer an increase in the firm's wealth against its generation of increasing flow of its profits.
- ☆ The market price of a share reflects the shareholder's expected return considering the long term prospects of the firm, reflects the differences in timings of the returns, considers risk and recognizes the importance of distribution or returns.

The maximization of a firm's value as reflected in the market price of a share is viewed as a proper goal of the firm. The profit maximization can be considered as a part of wealth maximization.

## 7.4.3 Organization of Finance Function

### 1. The Interface of Financial Policy with Corporate Strategic Management

The two important functions of the finance manager are:

- ☆ Allocation of funds (investment decision)
- ☆ Generation of funds (financing decisions)

The theory of finance makes two crucial assumptions to provide guidance to the finance managers in making these decisions. These are:

- ☆ The objective of the firm is to maximize the wealth of the shareholders.
- ☆ The capital markets are efficient.

The corporate finance theory implies that:

- ☆ Owners have the primary interest in the firm.
- ☆ The current value of share is the measure of shareholder's wealth.
- ☆ The firm should accept only those investments which generate positive net present values.
- ☆ The firm's capital structure and dividend decisions are irrelevant as they are solely guided by efficient capital markets and management no control over them.

However, the theory of finance has undergone fundamental changes over the past. It is felt that finance theory is not complete and meaningful without its linkage with the strategic management. Strategic management establishes an efficient and effective match between the firm's competence and opportunities with the risks created by the environmental changes.

**2. Interface of Finance Policy and Strategic Management**

Finance policy requires the resource deployment such as materials, labour etc. Strategic management considers all markets such as material, labour and capital as imperfect and changing. Strategies are developed to manage the business firms in uncertain and imperfect market conditions and environment. For forecasting, planning and formulation of financial policies, for generation and allocation of resources, the finance manager is required to analyze changing market conditions and environment.

The strategy focuses on how to compete in a particular product market segment or industry. For framing strategy it is considered that the shareholders are not the only interested group in the firm. There are many other influential constituents such as lenders, employees, customers, suppliers etc. The success of a company depends on its ability to survive in product market environment which is possible only when the company considered maintaining and improving its product market positions. Such considerations have important implications for framing corporate financial policies. Hence, the financial policy of a company is closely linked with its corporate strategy.

The company's strategy establishes an efficient and effective match between its competencies and opportunities and environmental risks. Financial policies of a company should be developed in the context of its corporate strategies. With the overall framework of the firm's strategy there should be a consistency between financial policies-investment, debt and dividend e.g. a company can sustain a high growth strategy only when investment projects generate high profits and it follows a policy of low payout and high debt.

**3. Interrelationship between Investment Financing and Dividend Decisions**

The finance functions are divided into 3 major decisions *viz.* investment, financing and dividend decisions. It is correct to say that these decisions are interrelated because the underlying objective of these 3 decisions is the same *i.e.* maximization of the shareholder's wealth. Since investment financing and dividend decisions are all interrelated one has to consider the joint impact of these decisions on the market price of the company's share and these decisions should also be solved jointly. The decision to invest in a new project needs finance for investment. The financing decision in turn is influenced by dividend decision because retained earnings used in internal financing deprive shareholder of their dividends. An efficient finance management can ensure optimal joint decisions. This is possible by evaluating each decision in relation to its effects on the shareholder's wealth.

The impact of taxation on corporate financial management

The tax payments represent a cash outflow from business and therefore these tax cash outflows are critical part of the financial decision making in business. Taxation affects a firm in numerous ways. The most significant effects are as under:

- ✰ Tax implication and financial planning: While considering the financial aspects or arranging the funds for carrying out the business, the tax implications arising there should also be taken into account. The income of the business undertakings is subject to tax at the rates given in finance act.
- ✰ The weighted average cost of capital is reduced because interest payments are allowable for computing taxable income.
- ✰ Where a segment of the firm incurs loss but the firm gets overall profits from other segments
- ✰ The income tax act allows depreciation on plant, furniture, and buildings owned by the assessed and used by him for carrying on his business, occupation and profession. This deprecation is allowed for full year if an asset was used for the purpose of business or profession for more than 180 days. Unabsorbed depreciation can be carried forward indefinitely.
- ✰ Capital budgeting decisions: The setting up of a new project involves consideration of tax effects. The decision to set up a project under a particular form of business organization at a particular place, choice of nature of business, and the type of activities to be undertaken etc. requires that a number of tax considerations should be taken into account before arriving at the appropriate decision from the angle of sound financial management. The choice of particular manufacturing activity may be influenced by the special tax concessions available such as-
    - Higher depreciation allowance
    - Amortization of expenditure on know-how, scientific research related to business, preliminary expenses etc.
    - Deductions in respect of profit derived from the publication of books etc.
    - Deductions in respect of profit derived from export business.

### 7.4.3 Financial Management and Accounting

Similar to production and sales, finance is an independent specialized function, integrated with other functions. Financial management is a separate management area. Many organisations have one department for accounting and finance functions and the finance function is often considered as part of the functions of the Accountant. But the Financial management is something more than an art of accounting and book keeping in the sense that, accounting function discharges the function of systematic recording of transactions relating to the firm's transactions in books of account and summarizing the same for presenting in financial statements *viz.*, profit and loss account and balance sheet, funds flow and cash flow statements.

The finance manager uses the accounting information in analysis and review of the firm's business position to make decisions. Besides the analysis of financial information available from the books of account and records of the firm, a finance manager can also use techniques like capital budgeting techniques, statistical and mathematical models and computer applications in decision making to maximize the value of the firm's wealth and value of the owners' wealth. Considering these facts, finance function is a distinct and separate function rather than simply an extension of accounting function. Financial management being an important function; many firms prefer to centralize the function to keep constant control on the finances of the firm. Any inefficiency in financial management will be disastrous for the firm, but for the routine matters, the finance function could be decentralized with adoption of responsibility accounting concept. It is advisable to decentralize accounting function to speed-up the process of information. But since the Recounting information is used in taking financial decisions, proper control should be exercised on accounting functions in processing of accurate and reliable information to the needs of the firm. The centralization or decentralization of accounting and finance functions mainly depends on the attitude of the top level management.

### 7.4.5 Significance of Financial Management

The importance of financial management can be understood from the following aspects:

- **Applicability:** The principles of finance are applicable wherever there is cash flow. The concept of cash flow is one of the central elements of financial analysis, planning control and resource allocation decisions. Cash flow is important because the financial health of the firm depends on its ability to generate sufficient amounts of cash to pay its employees, suppliers, creditors and owners. Any organisation, whether motivated with earning of profit or not, having cash flow requires to be viewed from the angle of financial discipline. Therefore, financial management is equally applicable to all forms of business like sole traders, partnerships and companies. It is also applicable to non-profit organizations like trusts, societies, government organisations, public sector enterprises etc.
- **Chances of Failure:** A firm having latest technology, sophisticated machinery, highly capable marketing and technical experts, etc. may fail unless its finances are managed on sound principles of Financial Management. The strength of business lies in its financial discipline. Therefore, finance function becomes primary, enabling the other functions like production, marketing, purchase, personnel etc. to be more effective in achievement of organizational goals and objectives.

Return on Investment Anybody who invests his money will earn a reasonable return on his investment. The owners of business try to maximize their wealth. It depends on the amount of cash flows expected to be generated for the benefit of owners, the timing of these cash flows and the risk attached to these cash flows. The greater the time and risk associated with the expected cash flow, the greater is the rate of return required by the owners. The Financial management studies the risk-return perception of the owners and the time value of money.

### 7.4.6 Strategic Financial Management

Strategic planning is long range in scope and has its focus on the organization as a whole. The concept is based on an objective and comprehensive assessment of the present situation of the organization and the setting up of targets to be achieved in the context of an intelligent and knowledgeable anticipation of changes in the environment. The strategic financial planning involves financial planning, financial forecasting, provision of finance and formulation of finance policies which should lead the firm's survival and success. The responsibility of a finance manager is to provide a basis and information of strategic positioning of the firm in the industry. The firm's strategic financial planning should be able to meet the challenges and competition and it would lead to firm's failure or success. The strategic financial planning should enable the firm to judicious allocation of funds, capitalization of relative strengths, mitigation of weaknesses, early identification of shifts in environment, counter possible actions of competitor, reduction in financing costs, effective use of funds deployed, timely estimation of funds requirement, identification of business and financial risk etc.

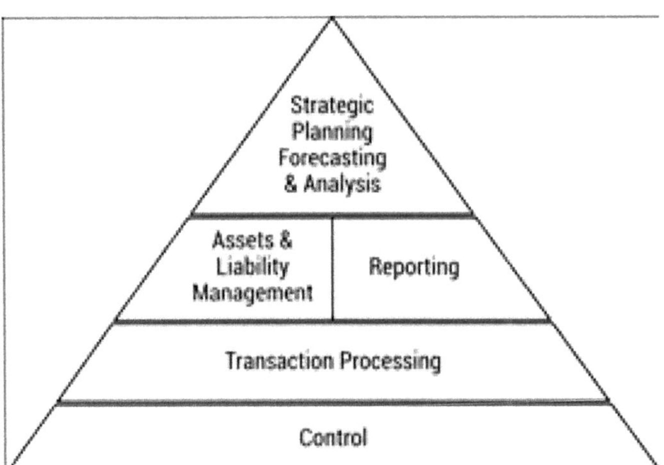


**Fig. 7.6: Elements of Strategic financial planning**

The strategic financial planning is needed to counter the uncertain and imperfect market conditions and highly competitive business environment. While framing financial strategy, the shareholders should be considered as one of the constituents of a group of stakeholder's *viz.*, shareholders, debenture holders, banks, financial institutions, government, managers, employees, suppliers, customers etc.

The strategic planning should concentrate on multidimensional objectives like profitability, expansion growth, survival, leadership, business success, positioning of the firm, reaching global markets, brand positioning etc. The financial policy requires the deployment of firm's resources for achieving the corporate strategic objectives. The financial policy should align with the company's strategic planning. It allows the firm in overcoming its weaknesses, enable to maximize the utilization of its competencies and mould the prospective business opportunities and threats to the advantage of the firm. Therefore, the finance manager should take the investment and finance decisions in consonant to the corporate strategy. In accordance with the classical management theory, the financial function of an enterprise has five main objectives, *viz.*, forecasting, organizing, planning, co-ordination and control. Each of these objectives has its own range of related themes.

**Forecasting**

- ☆ Demand and sales volume/revenues
- ☆ Cash flows
- ☆ Prices
- ☆ Inflation rates
- ☆ Labour union behaviour
- ☆ Technology changes
- ☆ Inventory requirements

**Organizing**

- ☆ Financial relation
- ☆ Liaison with financial institutions and clients
- ☆ Accounting system

**Planning**

- ☆ Investment planning
- ☆ Man power planning
- ☆ Development process
- ☆ Marketing strategies

**Co-ordination**

- ☆ Linking finance function with other areas
- ☆ Linking with national budget and five year plans
- ☆ Linking with labour union policies
- ☆ Liaison with media

**Control**

- ☆ Financial charges

- ☆ Achievement of desired objectives
- ☆ Overall monitoring of the system
- ☆ Equilibrium in the capital

# Questions

## Short Answer Questions

### Write short notes on

1. Maximization of Profit
2. Role of Finance manager
3. Objectives of financial Management
4. Equity Capital
5. Term Loans
6. Debentures

## Long Answer Questions

1. Explain the profit and wealth maximization concept. Which should be given the importance in the long run?
2. Write a detailed note on long term sources of finance.

# Glossary

**Account Receivable :** A balance due from a customer.

**Accounting Profit :** A firm's net income as reported on its income statement.

**Accruals :** Continually recurring short-term liabilities, especially accrued wages and accrued taxes.

Annual Report" A report issued annually by a corporation to its stockholders. It contains basic financial statements, as well as management's opinion of the past year's operations and the firm's future prospects.

**Balance Sheet :** Also known as the financial statement or statement of net worth, the balance sheet provides an overall financial snapshot of the agri-business business on a specific date. It lists all of the assets (property) and liabilities (loans) of the business as of the balance sheet date. Income and expense records are not needed for the balance sheet. A value for each asset and the outstanding balance for each liability is given in the balance sheet. Ideally, the balance sheet should separate assets and liabilities into current (less than one year of life), intermediate (one to seven years of life) and long term (longer than seven years of life, mainly buildings and land) categories and should list cost (original cost less depreciation) and market value (current expected sale price) for each.

**Cash Accounting System :** Cash accounting systems list income and expense items in a general ledger framework. Individual income and expense accounts are not used or reconciled. Enterprise analysis can be achieved, however, and inventory adjustments allow for accurate calculations of an accrual income statement. For accuracy, reconciling records back to a bank statements is important.

**Double-Entry Accounting System :** Double-entry accounting contains individual income and expense chart of accounts. Each income or expense entry is actually entered twice, reconciling individual accounts back to the general ledger. While accurate and detailed, these systems can be tedious and time-consuming to maintain.

**Enterprise Accounting :** Enterprise accounting requires that income and expense information be assigned to the agri-business enterprise that generated that income or expense. "Enterprise," as used in this publication, refers to the different kinds of agri-business production. For example, a agri-business may have a corn enterprise and a soybean enterprise. Enterprises may be divided within one single commodity, such as dividing a corn enterprise between no-tillage and conventional tillage fields.

**Expenses :** These items include direct production expenses, fixed or overhead expenses, capital expenditures and personal and family living expenses. Basically, expenses refer to any and all money spent. Direct production expenses are also referred to as variable expenses because they vary with the amount and level of production. Feed, seed and fertilizer expenses are examples of costs that will vary as the production program changes. Fixed expenses have to be made each year regardless of the production level. That is, these expenses will be incurred whether anything is produced or not. Such items include taxes, insurance, interest and rent. Capital expenditures are for items that have a useful life of more than one year. Such items include machinery purchases, breeding stock, facilities and structures, equipment and land improvements. Personal expenditures may also be referred to as family living expenses. These items, such as food, clothing, donations, etc., are not necessarily related to the agri-business operation, but do have a direct impact on the financial condition of the agri-business.

**Income :** This term means money received for selling product or service(s). It includes sales of purchased, breeding or raised livestock. It also includes sales of crops, government program proceeds and proceeds from cost-share projects. Custom work by the agri-business owner and family members would also be included as income. In short, all income that is generated by the agri-business or agri-business family is included as income.

**Income Statement :** Also known as the profit and loss statement, the income statement lists all expenses and income for the agri-business during an accounting period. In addition, the income statement accounts for changes in certain inventory items. For example, for a cow/calf operation, if heifers were held in the herd for breeding, cash income would not reflect their worth. But the change in the value of the cow herd would reflect their worth. This change in the inventory would be included on the income statement as agri-business revenue. Income, direct expenses, depreciation expenses, gains or loss on liquidated assets and non-agri-business income and expenses are included in the income statement.

**Net Worth :** The net worth is calculated on the balance sheet. It is the value of everything a business owns, less any loans or liabilities the business incurs. In essence, the net worth is the value of total agri-business assets minus total agri-business liabilities on a given date.

**Projected Monthly Cash Flow Statement :** The projected cash flow statement indicates the source and amount of income and expense activities for a given period. It also shows when money will be borrowed and when it will be repaid. The cash flow demonstrates the ability to repay a loan in a timely manner.

# Unit 8

# Working Capital Management in Agri-business

## 8.1 Introduction

Working capital is significant in financial management due to the fact that it plays an important role in keeping the wheels of a agri-business enterprise running. It may be regarded as lifeblood of a agri-business. It is concerned with short term financial decisions. Its effective provision can do much to ensure the success of a agri-business, while its inefficient management can lead not only to loss of profits but also to the ultimate downfall of promising concept. A study of working capital is of major importance to internal and external analysis because of its close relationship with day-to-day operations of a agri-business.

Working capital management includes management of various components of current assets as well as current liabilities. A firm invests a part of its permanent capital in fixed assets and keeps a part of it for working capital *i.e.* for meeting the day to day requirements. The requirements of working capital varies from firm to firm depending upon the nature of agri-business, production policy, market conditions, seasonality of operations, conditions of supply etc. Working capital management if carried out effectively, efficiently and consistently will assure the health of an organization.

## 8.2 Meaning and Definition

In accounting, W. C. is the difference between the inflow and outflow of funds. It is the net cash inflow.

W.C. is defined as the excess of current assets over its current liabilities and provision. Current assets are those assets which will be converted into cash with the current account period or within the next year as a result of the ordinary operations of the agri-business.

Efficient working capital management requires that the firms should operate with some amount of Net working capital (NWC), the exact amount varying from firm to firm and depending, among other things, on the nature of industry. The greater the margin by which the current assets cover the short term obligations, the more able it will be to pay its obligations when they become due for payment.

## 8.3 Types of Working Capital

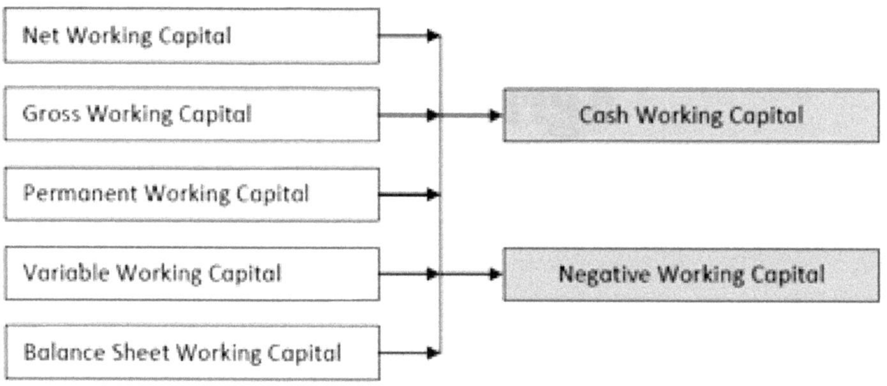


**Fig. 8.1: Type of Capital**

### 1. Networking Capital

Networking Capital is the difference between current assets and current liabilities. The concept of net working capital enables a firm to determine how much amount is left for operational requirements.

### 2. Gross Working Capital

Gross working capital is the amount of funds invested in the various components

- ☆ Financial Managers are profoundly concerned with current assets:
- ☆ Gross working Capital provides the current amount of working capital at the right time;
- ☆ It enables a firm to realize the greatest return on its investment;
- ☆ It helps in the fixation of various areas of financial responsibility;
- ☆ It enables a firm to plan and control funds and to maximize the return on investment.

For these advantages, gross working capital has become a more acceptable concept in financial management.

## 3. Permanent Working Capital

Permanent Working Capital is the minimum amount of current assets which is needed to conduct a business even during the dullest season of the year. This amount varies from year to year, depending upon the growth of a company and the stage of the business cycle in which it operates. It is the amount of funds required to produce the goods and services which are necessary to satisfy demand at a particular point. It represents the current assets which are required on a continuing basis over the entire year. It is maintained as the medium to carry on operations at any time. Permanent working capital has the following characteristics:

- ☆ It is classified on the basis of the time factors;
- ☆ It constantly charges from one asset to another and continues to remain in the business process
- ☆ Its size increase with the growth of business operations

## 4. Temporary or Variable Working Capital

It represents the additional assets which are required at different times during the operating year-additional inventory, extra cash, etc. Seasonal working capital is the additional amount of current assets- particularly cash, receivable and inventory which is required during the more active business seasons of the year. It is temporarily invested in current assets and possesses the following characteristics;

- ☆ It is not always gainfully employed, though it may change from one asset to another, as permanent working capital does;
- ☆ It is particularly suited to business of a seasonal or cyclical nature.

## 5. Balance Sheet Working Capital

The balance sheet working capital is not which is calculated from the items appearing in the balance sheet. Gross working capital which is represented by the excess of current assets, and net working capital which is represented by the excess of current assets over current liabilities are examples of the balance sheet working capital.

## 6. Cash Working Capital

Cash working capital is one which is calculated from the items appearing in the profit and loss accounts of current assets. This concept has the following advantages -

It shows the real flow of money or value at a particular time and is concerned to be the most realistic approach in working capital management. It is the basis of the operation cycle concept which has assumed a great importance in financial management in recent years. The reason is that the cash working capital indicates the adequacy of the cash flow which is an essential pre - requisite of a agri-business.

### 7. Negative Working Capital

Negative working Capital emerges when current liabilities exceed current assets. Such a situation is not absolutely theoretical and occurs when a firm is nearing a crisis of some magnitude.

## 8.4 Factors Determining Working Capital

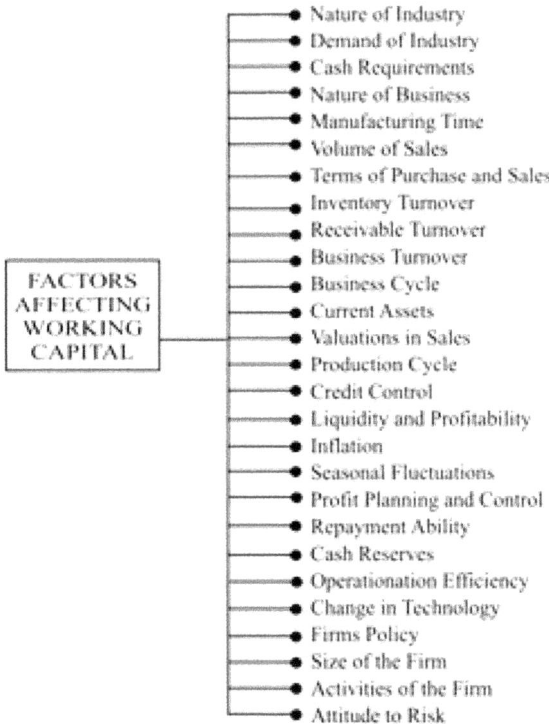


**Fig. 8.2: Factors Affecting Working Capital**

Railroad, with their large fixed investments, appear to have the lowest requirements for current assets. This does not mean that the problems of working capital may be minimized in this field of enterprise, since ready funds are still essential to cover disbursement for wages, interest on funded debt, purchase of materials and supplies, etc. Indeed, under such conditions the working capital position may become even more strategic in character because of its relation to, and control of, the large amount of fixed assets. Thus, one of the outstanding problems of railroad management in recent years has been the maintenance of a current position sufficiently strong to permit vigorous operations.

Public utilities like rail road's have large fixed investments which cause the current assets to constitute only a relatively small percentage of the total assets. There is a difference between operating and holding companies, but even then the funds to cover current transactions are minor as compared with those necessary to finance the long term structure.

Industrial concerns, generally, require a large amount of working capital, although it varies from business to business with lack of uniformity characterising each field of enterprise. However, the underlying determinants of the amount are essentially the same as in the earlier groups. Where large amounts of fixed capital are required for operation, working assets may be expected to occupy a smaller niche in the asset structure. For similar reasons, a rapid turnover of capital (sales divided by total assets) will inevitably means a large proportion of current assets in the case of industries with large fixed investment. One of the primary uses of working capital is its conversion into operation which will normally replace such defections. This means that the flow of a portion of the working capitals circulated through fixed investment and that its recovery is dependent upon the income realized. Where the current assets are relatively more important, a rapid sales turnover is usually found. Often, as in the case of retail concerns, the specific working assets constitute the object of sale and recovery is direct and immediate.

In agri-business, a large share of working capital is more likely to get converted into finished products, but even here, the potentiality of recovery is not delayed as much as in the case of public utilities and railroads. The need for working capital varies with changes in the volume of business. A considerable proportion of current assets are needed permanently as fixed assets. At the same time, new receivable accumulate and old ones are converted into cash. Cash is utilized in the production process.

The following factors determine the amount of working capital;

- ✰ **Nature of Industry:** The composition of an asset is a function of the size of a business and the industry to which it belongs. Small agri-business have smaller proportions of cash, receivables and inventory than large corporations. This difference becomes more marked in large corporations. A public utility, for example, mostly employs fixed assets in its operations, while merchandising department depends generally on inventory and receivables. Needs for working capital are thus determined by the nature of an enterprise.
- ✰ **Demand of Industry:** Creditors are interested in the security of loans. They want their obligations to be sufficiently covered. They want the amount of security in assets which are greater than the liability.
- ✰ **Cash Requirements:** Cash is one of the current assets which is essential for successful operations of the production cycle. Cash should be adequate and properly utilised. It would be wasteful to hold excessive cash. A minimum level of cash is always required to maintain good credit good credit relations. Richards Osborn has pointed out that cash has a universal liquidity and acceptability. Unlike illiquid assets, its value is clear-cut and definite.
- ✰ **Nature of Agri-business:** The nature of agri-business is an important determinant of the level of the working capital. Working capital requirements depend upon the general nature or type of business. They are relatively low in public utility concerns, in which inventories and receivables are rapidly converted into cash. Manufacturing organizations, however, face problems

of slow turnover of inventories and receivables, and invest large amount in working capital.

- **Time:** The level of working capital depends upon the time required to manufacture goods. If the time is longer, the size of working capital is greater. Moreover, the amount of working capital depends upon inventory turnover and the unit cost of the goods that are sold. The greater this cost, the bigger is the amount of working capital.
- **Volume of Sales:** This is the most important factor affecting the size and components of working capital. A firm maintains current assets because they are needed to support the operational activities which result in sales. The volume of sales and size of the working capital are directly related to each other. As the volume of sales increases, there is an increase in the investment of working capital in the cost of operations, in inventories and in receivables.
- **Terms of Purchase and Sales:** If the credit terms of purchase are more favourable and those of sales less liberal, less cash will be invested in inventory. With more favourable credit terms, working capital requirements can be reduced. A firm gets more time for payment to creditors or suppliers. A firm which enjoys greater credit with banks needs less working capital.
- **Inventory Turnover:** If the inventory turnover is high, the working capital requirements will be low. With a better inventory control, a firm is able to reduce its working capital requirements. While attempting this, it should determine the minimum level of stock which it will have to maintain throughout the period of its operations.
- **Receivable Turnover:** It is necessary to have an effective control of receivables. A prompt collection of receivables and good facilities for settling payables result into low working capital requirements.
- **Agri-business Turnover:** The business turnover of the organization directly calls for systematic planning for production. The exploitation of the available business can be achieved only when sufficient raw materials are stored and supplied. Hence Agri-business Turnover will also influence the working capital.
- **Agri-business Cycle:** Agri-business expands during periods of prosperity and declines during the period of depression. Consequently, more working capital is required during periods of prosperity and less during the periods of depression. During marked upswings of activity, there is usually a need for larger amounts of capital to cover the lag between collection and increased sales and to finance purchases of additional materials to support growing business activity. Moreover, during the recovery and prosperity phase of the business cycle, prices of raw materials and wages tend to rise, requiring additional funds to carry even the same physical volume of business. In the downswing of the cycle, there may be a brief period when collection difficulties and declining sales together cause embarrassment by requiring replenishing of cash. Later, as the depression runs its course, the concern

may find that it has a larger amount of working capital on hand than current business volume may justify.

- **Volume of Current Assets:** A decrease in the real value of current assets as compared to their book value reduces the size of the working capital. If the real value of current assets increases, there is an increase in working capital.
- **Variation of Sales:** A seasonal business requires the maximum amount of working capital for a relatively short period of time.
- **Production Cycle:** The time taken to convert raw materials into finished products is referred to as the production cycle or operating cycle. The longer the production cycle, the greater is the requirement of working capital. An utmost care should be taken to shorten the period of the production cycle in order to minimize working capital requirements.
- **Credit Controls:** Credit Controls includes such factors as the volume of credit sales, the terms of credit sales, the collection policy, etc. With a sound credit control policy, it is possible for a firm to improve its cash inflow.
- **Liquidity and Profitability:** If a firm desires to take a greater risk for bigger gains or losses, it reduces the size of its working capital in relation to its sales. If it is interested in improving its liquidity, it increases the level of its working capital. However, this policy is likely to result in a reduction of the sale volume, and, therefore, of profitability. A firm, therefore, should choose between liquidity and profitability and decide about its working capital requirements accordingly.
- **Inflation:** As a result of inflation, size of the working capital is increased in order to make it easier for a firm to achieve a better cash inflow. To some extent, this factor may be compensated by the rise in selling price during inflation.
- **Seasonal Fluctuations:** Seasonal Fluctuations in sales affect the level of variable working capital. Often, the demand for product may be of a seasonal nature. Yet inventories have got to be purchased during certain season only. The size of the working capital in one period may, therefore, be bigger than that in another.
- **Profit Planning and Control:** The level of working capital is decided by the management in accordance with its policy of profit planning and control. Adequate profit assists in the generation of cash. It makes it possible for the management to plough back a part of its earnings in the business and substantially build up internal financial resources. A firm has to plan for taxation payments, which are an important part of working capital management. Often dividend policy of a corporation may depend upon the amount of cash available to it.
- **Repayable Ability:** A firm's repayment ability determines level of its working capital. The usual practice of a firm is to prepare cash flow projections according to its plans of repayment and to fix working capital levels accordingly.

- **Cash Reserves:** It would be necessary for a firm to maintain some cash reserves to enable it to meet contingent disbursements. This would provide a buffer against shortages in cash flows.
- **Operational and Financial Efficiency:** Working capital turnover is improved with a better operational and financial efficiency of a firm. With a greater working capital turnover, it may be able to reduce its working capital requirements.
- **Changes in Technology:** Technology developments related to the production-process have a sharp impact on the need for working capital.
- **Firms Policies:** These affect the level of permanent and variable working capital. Changes in credit policy, production policy, etc., are bound to affect the size of working capital.
- **Size of the Firm:** A firm's size, either in terms of its assets or sales, affects its need for working capital. Bigger firms, with many sources of funds, may need less working capital as compared to their total assets or sales.
- **Activities of Agri-business:** Agri-business stocking on heavy inventory or selling on easy credit terms calls for a higher level of working capital for it than for selling services or making cash sales.
- **Attitude of Risk:** The greater the amount of working capital, the lower is the risk of liquidity.
- Whenever there is current strain, it has to be immediately diagnosed on the basis of the red signals which manifest themselves in the operation. The cause should be ascertained by making thorough study of the components of current assets and current liabilities.

If stock is not moving fast, and if there is an excess inventory build-up, corrective steps should be taken to sell the stock or bring down its level. If the receivables have become sticky, effective recovery steps should be taken to reduce the debts and to increase the collections. Sometimes short-term funds have been used to finance fixed assets, and this creates the current strain. This imbalance in the pattern of financing should be set right by raising long-term funds on the cover of fixed assets so that the current strain may be wiped out. Similarly, if current funds are diverted outside when they are badly required within the firm itself, it would be very difficult to run the agri-business. External diversion may be for the purpose of outside investment, advance to other or allied concerns or may be in the form of drawing from the business or for various other purposes.

The situation can improve only if this external diversion is stopped. If the strain is allowed to continue agri-business of involvement in any other business or industry, the consequences may be disastrous. In such a situation, the ability to meet current demands deteriorates; short-term credits are not forthcoming; production is affected; sales decline; cash flow dwindles; income may disappear; and the whole enterprise may get into the red over a period of time.

## 8.5 Operating Working Capital Cycle

A new concept which is gaining more importance in recent years is the operating Cycle concept' of Working Capital. The operating cycle refers to the average time elapses between the acquisition of raw materials and the final cash realization. Then the raw materials and stores are issued to the production department. Wages are paid and other expenses are incurred in the process and work-in-process comes into existence.

Work-in-process becomes finished goods. Finished goods are sold to customers on credit. In the course of time, these customers pay cash for the goods purchased by them. Cash is retrieved and the cycle is completed. Thus, operating cycle consists of four stages:

- ☆ The raw materials and stores inventory stage
- ☆ The work- in- progress stage
- ☆ The finished goods inventory stage
- ☆ The receivable stage

The operating cycle begins with the arrival of the stock, and ends when the cash is received. The cash cycle begins when cash is paid for materials, and ends when cash is collected from receivables.

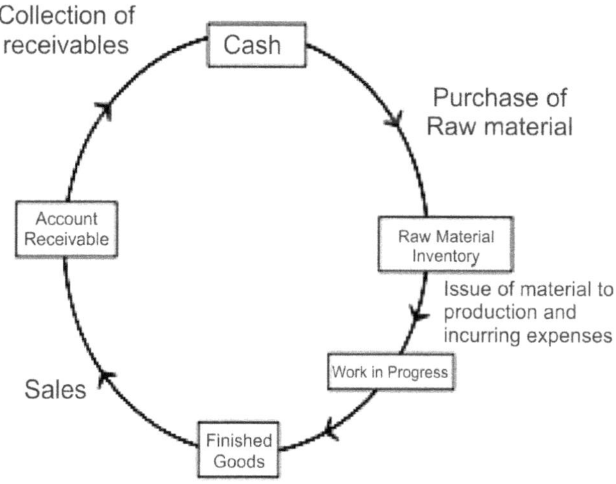


**Fig. 8.3: Operating working Capital Cycle**

**Importance of Operating Cycle Concept :** The application of operating cycle concept is mainly useful to ascertain the requirement of cash working capital to meet the operating expense of a going concern. This concept is based on the continuity of the flow of value in a business operation.

This is an important concept because the longer the operating cycle, the more working capital funds the firm needs. Management must ensure that this cycle does not become too long. This concept precisely measures the working capital fund requirements, traces its changes and determines the optimum level of working capital requirements.

## 8.6 Working Capital Requirements

Every agri-business firm requires at least some amount of working capital, only their working capital requirements are different. The goal of every agri-business firm should be to increase the profit of its shareholders. And to achieve this goal, the firm's operations must yield enough returns. Successful sales activity is very essential to earn profit and hence, it is imperative that a firm invests sufficient funds in its current assets for sales generation. As sales cannot be converted into cash immediately, current assets are required to convert sales into cash.

Working capital is essentially circulating capital; in fact it is often referred to as such. This has been admirably summed up by comparing it with arriver which is there every day, but the water in it is constantly changing.

The required amount of working capital in relation to the fixed capital of agri-business will vary widely between firms in different industries. For example, an agri-business company engaged in agriculture will need a large amount of fixed or long term capital to finance the agri-business, equipment, etc for considerable periods, whereas a jobbing builder will require virtually no fixed assets but instead a reasonably large amount of working capital to finance stocks of parts, amounts owing by customers, etc. If company does not have enough working capital it will soon find its activities restricted. Many firms which seemed to be expanding their activities successfully have faced trouble through insufficient working capital being available to finance this expansion. Under normal conditions a steady increase in working capital indicates a successful agri-business, while a steady decrease would be a danger signal demanding immediate action to remedy the situation. Both in practice and in examinations, the question is often asked:

What will be the working capital requirements to finance this level of activity or that new project? This is a very practical and important problem which may require extensive research and difficult calculations. However, to show the usual requirements in simple form, the following items are tabulated:

- ☆ The cost of raw materials, wages and overheads.
- ☆ The period during which raw materials will remain in stock before issue to production.
- ☆ The period during which the product will be processed through the factory.
- ☆ The period during which finished goods will remain in the warehouse.
- ☆ The lag in payment to suppliers of raw materials and service.
- ☆ The lag in payment to employees.
- ☆ The lag in payment by debtors.
- ☆ Frequently an amount is allowed to cover contingencies, e.g. 10% might be added to the total amount.

Most of these points will be included in a computation of working capital requirements. For example, the period during which stocks of finished goods stay in the warehouse can only be an average figure, but by careful observation it should be possible to make a reasonable assessment. This is the reason for allowing an amount to cover contingencies: it is hoped that this figure might cover any inaccuracies in calculation.

## 8.7 Estimating Working Capital Needs and Financing Current Assets

Operating cycle is the most appropriate method for computing the working capital requirements of any company. However, methods other than operating cycle for computing a firm's working capital include:

- Estimation of a agri-business firm's working capital requirements on the grounds of its current assets' average holding period and then relating them to the company's costs based on past experiences. Operating cycle approach forms the basis of this method. In order to use this method for estimating working capital needs, one can make certain assumptions such as there is a supply of raw materials and semi finished and finished goods for one month.
- Assuming that the current assets vary with sales, the ratio of sales can be used as a method for estimating the working capital of a firm. Here, it can be assumed that the annual sales are anywhere between 25-30 %.
- Working capital requirement can also be estimated as a ratio of fixed investments.

One can assume the agri-business firm's capital investment to be 10-20% in order to use this method. As the second approach is clearly dependant on how accurately the sales are estimated, this method is less reliable. Likewise, the third method is highly dependent on the investment estimates. If the investments are not estimated properly, then it will affect the estimation of working capital needs using this method and hence, this method is not used generally.

Various factors like the accurate sales and investment forecasting, alterations in operations etc. should be considered while estimating a firm's working capital requirements. Also, other factors like the company's production cycle, its collection policies etc. should also be taken into consideration.

### Financing Current Assets

The various policies for financing current assets include:

- Long-term financing can be obtained by means of debentures, ordinary as well as preference share capital, long term loans coming from banks and financial institutions etc.
- Short-term financing can be sourced in the form of working capital coming from banks, commercial papers, public deposits etc. Short-term finance is obtained from short-term suppliers in money market as well as from banks and is generally for a period less than one year.

The automatic funds arising in day-to-day business are referred to as Spontaneous financing. Outstanding expenses, trade credit etc. are some of the examples of spontaneous financing. A firm is expected to use this source of financing to the fullest extent as there are no explicit costs associated with this type of financing.

## 8.8 Inventory Management

Inventories are asset items held for sale in the ordinary course of business or goods that will be used or consumed in the production of goods to be sold. The description and measurement of inventory require careful attention because the investment in inventories is frequently the largest current asset of merchandising (retail) and manufacturing businesses.

An inventory can be defined as The raw materials, work-in-process goods and completely finished goods that are considered to be the portion of a business's assets that are ready or will be ready for sale. Inventory represents one of the most important assets that most businesses possess, because the turnover of inventory represents one of the primary sources of revenue generation and subsequent earnings for the company's shareholders/owners.

Keeping the stock of inventories involves locking of the company's funds and increase in storage and handling costs. Companies hold inventories basically for 3 motives:

- Transactions motive *i.e.* to ensure smooth production and sales operations.
- Precautionary motive *i.e.* to guard against the volatility of demand and supply factors.
- Speculative motive which is done to take benefit of price fluctuations.

### 8.8.1 Types of Inventories

The various forms of inventories include

- Raw material inventories are the ones that form the basic ingredients to be converted into finished products by means of manufacturing process. A company buys and stores the raw materials to be used later.
- Work-in-process inventories are semi-finished goods and require more processing before becoming finished goods that are ready for sale.
- Finished goods inventories are completely ready products that can be sold.
- Other than above three basic inventories, there is one more inventory type called as supplies, also known as stores and spares. Though they are not directly into production, but are required for the process of production. Examples of supplies are materials like soaps, brooms, oil, fuel, light etc.

Inventory Management aims at reducing the direct and indirect costs that are associated with the inventory. It also includes maintaining an appropriate and sufficient inventory supply for the smooth production and sales activities. The severity of inventory management of a firm depends upon the extent to which the firm has invested in inventories. An ideal and efficient inventory management -

- Must ensure that the raw materials are supplied at the right time and quantities for smooth production operations.
- Should anticipate price fluctuations and accordingly keep enough supply of raw materials in times of shortage.
- Maintain a healthy supply of finished goods for better customer service as well as smooth sales activities
- Reduce the carrying time and expenses
- Exercise control over investments made in inventories.

### 8.8.2 Inventory Management Techniques

To determine the optimum level of inventory, there should be proper inventory management techniques which in turn should be in accordance with the shareholders'wealth maximization. The importance of effective inventory management depends on the size of the investment of the inventory. A firm should use a systematic approach to manage its inventory. There are 3 different inventory management techniques.

- Economic Order Quantity (EOQ)
- Reorder Point
- Stock Level

### 8.8.3 Pricing of Inventories

Following are the different ways of valuing the inventories and these methods are important to have an efficient inventory management process. These methods are used to value the raw materials:

- First-in-first-out (FIFO)
- Last-in-first-out(LIFO)
- Weighted Average Cost method
- Standard Price method
- Replacement or Current Price method

## Questions

### Short Answer Questions

1. Importance of capital budgeting
2. Superiority of cash flow concept over accounting profit concept
3. Comparison of NPV and the Internal Rate of Return (IRR)
4. Purposes of holding the inventories
5. Types of inventories

## Long Answer Questions

1. Discuss the various ways through which the working capital cycle can be shorten.
2. What are the factors for efficient cash management?
3. Explain the role of capital budgeting techniques in investment making decisions.
4. What are types of Capital Structure?
5. What are the techniques of Inventory Management?
6. Discuss cost of maintaining receivables

# Glossary

**Working Capital -** A firm's investment in short-term assets--cash, marketable securities, inventory, and accounts receivable.

**Working Capital Policy -** Basic policy decision regarding (1) target levels for each category of current assets and (2) how current assets will be financed.

# Unit 9

# Agribusiness Capital Budgeting

## 9.1 Introduction

The term Capital budgeting contains the relatively scarce, non-human resource of production enterprise, and budgeting, indicating a detailed quantified planning which guides future activities of an agri-business enterprise towards the achievement of its profit goals. Capital refers to total funds employed in an agri-business enterprise as a whole.

## 9.2 Meaning of Capital Budgeting

The Capital fund is increased by an inward flow of cash and decreased by an outward flow of cash and as such it is important for an enterprise to plan and arrange cash flows properly.

Capital budgeting, then, consists in planning the development of available capital for the purpose of maximizing the long-term profitability.

Capital budgeting may be defined as the decision-making process by which a agri-business firm evaluates the purchase of major fixed assets, including buildings, machinery, and equipment. It deals exclusively with major investment proposals which are essentially long-term projects and is concerned with the allocation of firm's scarce financial resources among the available market opportunities, a search for a new and more profitable investment proposal and the making of an economic analysis to determine the profit potential of each investment proposal. They are large, permanent commitments which influence its long-run flexibility and earning power.

It is a process by which available cash and credit resources are allocated among competitive long-term investment opportunities so as to promote the greatest profitability of agri-business company over a period of time. It refers to the total process of generating, evaluating, selecting and following up on capital expenditure alternatives.

## 9.3 Principal of Capital Budgeting

Capital expenditure decision should be taken on the basis of following factors-

- ☆ Creative search for Profitable opportunities-The first stage is the conception of profit making idea. Profitable investment opportunities should be sought to supplement existing proposals.
- ☆ Long range capital planning-A flexible programme of a company's expected future development over a long period of time should be prepared.
- ☆ Short range capital planning -This is for short period. It indicates its sectoral demand for funds to stimulate alternative proposals before the aggregate demand for funds is available.
- ☆ Measurement of project work-The economic worth of a project to a company is evaluated at this stage. The project is ranked with other projects.
- ☆ Screening and Selection-The project is examined on the basis of selection criteria, such as supply and cost of capital, expected returns, alternative investment opportunities etc.
- ☆ Control of authorized outlays-Outlay should be controlled in order to avoid costly delays and cost over-runs.
- ☆ Post Mortem-The ex-post routines of a completed investment project should be re-evaluated in order to verify their exact conformity with exact projections.
- ☆ Forms and Procedures-These involve the preparation of reports necessary for any capital expenditure programme.
- ☆ Economics of capital budgeting-It includes estimating the rate of return on capital expenditure. Knowledge of economic theory underlying investment decisions is needed for this purpose.

## 9.4 Kinds of Capital Budgeting Proposals

- ☆ Replacement
- ☆ Expansion
- ☆ Modernization of Investment Expenditures
- ☆ Strategic Investment Proposals
- ☆ Diversification
- ☆ Research and Development

Capital budgeting refers to the total process of generating, evaluating, selecting and following upon capital expenditure alternatives. The firm allocates or budgets financial resources to new investment proposals. Basically the firm may be confronted with three types of capital budgeting decisions:

- ✰ Accept-reject decisions
- ✰ Mutually Exclusive Project decisions
- ✰ Capital rationing decisions.

## 9.5 Capital Budgeting Techniques

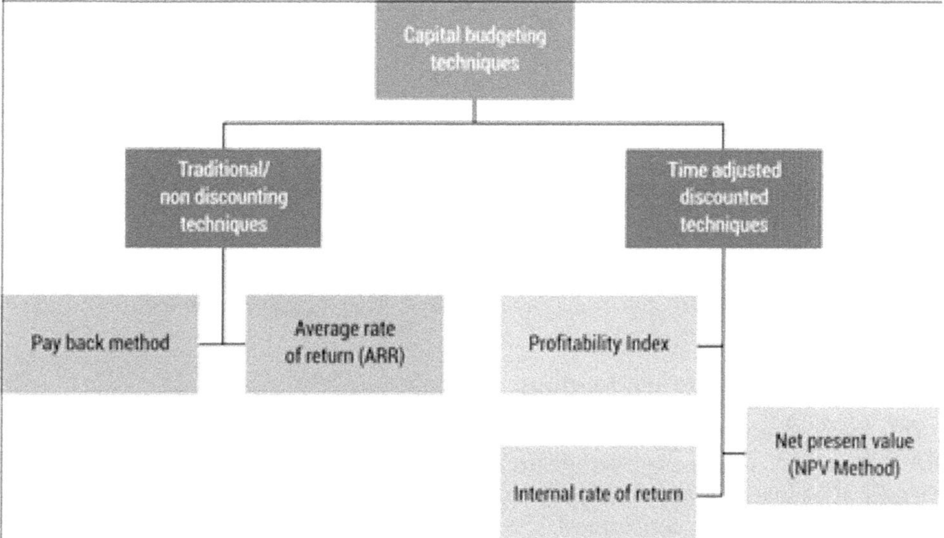


**Fig 9.1 Capital Budgeting Techniques**

### 9.5.1 Pay Back Method

The pay back method (PB) is the traditional method of capital budgeting. It is the simplest and perhaps, the most widely employed quantitative method for appraising capital expenditure decisions. This method answers the question - How many years will it take for the cash benefits to pay the original cost of an investment? Cash benefits here represent cash flows after taxes ignoring interest payment. This method measures the number of years required for the CFAT to pay back the original outlay required in an investment proposal.

There are two ways of calculating the PB period. The first method can be applied when the cash flow stream is in the nature of annuity for each year of the project's life *i.e.* CFAT are uniform.

PB Period = Investment

Constant annual cash flow

The second method is used when a project's cash flows are not equal, but vary from year to year. In such a situation, PB is calculated by the process of cumulative cash flows till the time when cumulative cash flows become equal to the original investment outlay.

If the actual payback period is less than the pre-determined pay back, the project would be accepted; if not, it would be rejected. When mutually exclusive projects are under consideration, they may be ranked according to the length of the payback period. Thus, the agri-business project having the shortest pay back may be assigned rank one.

## 9.5.2 Discounted pay-back period (DPP)

In this method the cash inflows are discounted at a rate which is equal to cost of capital and then payback period is worked out. This is better than ordinary payback period method as DPP considers the time value of money.

DPP = Investment

Discounted Annual Cash Inflow

The DPP is expressed in years and as long as the DPP is lesser than the estimated life of the project, the project is economically feasible and it can be accepted. Sometimes the two projects may have the same DPP although the estimated life of the project may be different.

### Average Rate of Return Method

This method is also known as Accounting Rate of Return. This method considers A.R.R. which means the average annual yield of the project. Under this method profit after tax and depreciation (also called as accounting profit) of a percentage of total investment is considered.

The annual returns of a project are expressed as a percentage of the net investment in the project. This method consists of aggregating all the earnings after depreciation and dividing them by the profits in useful lifespan. The resultant average earnings over the period is divided by the average investment over the period. The average investment in a agri-business project is always ½ of the original investment.

## 9.5 3 Internal Rate of Return Method (IRR)

The IRR Method is yet another discounted cash flow technique which takes into consideration the magnitude and timing of cash flows. It is also known as time adjusted rate of return, marginal efficiency of capital, marginal productivity of capital, and yield on investment and so on. It is employed with the cost of investment and the annual cash inflows are known while the unknown rate of earnings is to be ascertained.

The Internal Rate of Return (IRR) is that rate at which the sum of discounted cash inflows equals the sum of discounted cash out flows. It is the rate at which the NPV (Net Present Value) of the investment is zero. It is called internal rate because it depends mainly on the outlay and proceeds associated with the investment and not on any rate determined outside the investment.

**Advantages**

- ✰ The IRR method considers the time value of money like the NPV method.
- ✰ It takes into consideration the cash flows over the entire lifespan of the project.
- ✰ Business executives and non-technical people understand the concept of Internal Rate of Return (IRR) much better than that of NPV.

## 9.6 Estimation of Cash Flow for New Project

Evaluation of a project should rather be based on cash flows as distinct from accounting profits.

Accounting profits are useful only for external reporting purposes and by them cannot be used for the purpose of capital budgeting decisions. Accounting profits are capable of distortion because of possible bias of individual accountants. Different methods of providing depreciation and valuing closing stocks can lead to different accounting figures. On the other hand, cash flows are totally independent of different methods of accounting for depreciation and stocks. Properly calculated cash flows cannot be challenged by anyone.

Accountants do start with rupee coming in and rupee going out when they write cash book. But when calculating accounting income they carry out certain adjustments for depreciation, accruals and stock valuations. It is not always easy to convert the customary accounting profits back into actual cash flows but this has to be done.

We had seen earlier that cash flow = PAT + DEP but it is not always that simple in actual practice. Some basic principles of cash flows which are:

- ✰ Cash flows should be estimated on an incremental basis. A true worth of agri-business project depends only on additional cash flows that can be generated if a project is accepted.
- ✰ It is necessary to take into consideration all incidental effects of agri-business project on remainder of the business. Introduction of a new product may affect revenue of an existing product and therefore we must take into consideration this fact which is not normally recordable in accounting.
- ✰ Sunk costs should be ignored because they do not affect outflow of cash now. Whether we accept a project or reject it, sunk costs do not change and should therefore be ignored.
- ✰ Opportunity costs cannot be ignored. This was hinted at point number two above. This can be stretched a little further. Suppose a new project starts on a piece of land, which is not now acquired. However, if it is decided to sell the land for a sum and if it is not used for the new project then there is an opportunity cost or an implied outflow. Sale proceeds of the land will have to be sacrificed to implement the new agri-business project.
- ✰ Any new project will always entail an additional investment in short term assets like cash, debtors and stocks. The additional investment to some

extent is reduced to the extent to which short-term liabilities like creditors increase. The increase in net working capital requirements becomes an outflow and when the project finally ends, investment in working capital can be recovered either fully or partly. This will become an inflow.

- Cash flows should always be estimated on an after tax basis.
- Inflation, if persisting should always be treated on a consistent basis. This requires conversion of nominal cash flows into real cash flows.

## 9.7 Sources of Long Term Funds

Tabular Presentation of Sources of Long Term Funds in agri-business

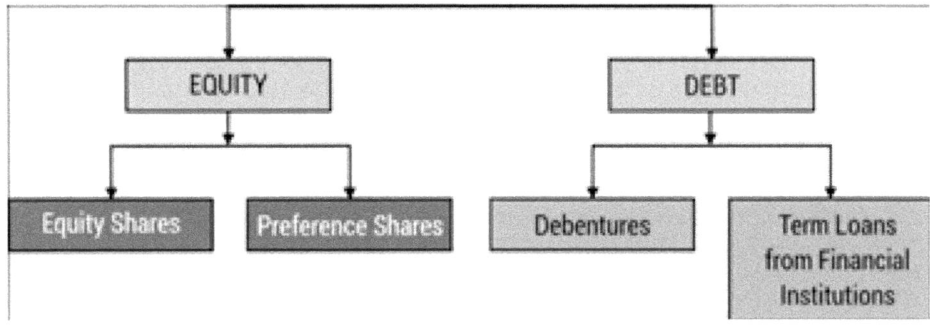


**Fig. 9.2: Source of long term fund**

Brief Characteristics:

- Interest on debentures or Term loans has to be paid whether there is any profit or not.
- Interest is a tax-deductible expense
- Dividend on preference shares is paid at fixed rate only if there is adequate profit after tax. If preference shares are cumulative then the dividends not paid will accumulate and will become payable in future.
- On equity shares, there is no fixed rate of dividend. Dividend may be skipped if profits are inadequate. Dividend may be very high if there is a bumper profit. Dividend can only be paid after preference dividend (including arrears if any) are paid and transfer to General Reserve Debenture and Redemption Reserve are made.

## Questions

### Short Answer Questions

1. Types of inventories in agri-business
2. Economic Order Quantity (EOQ)
3. Working Capital required for agri-business businesses
4. Disadvantages of inadequate working capital for agri-business

## Long Answer Questions

1. Explain the role of capital budgeting techniques in investment making decisions.
2. What are types of Capital Structure of agri-business?
3. What are the techniques of Inventory Management?
4. Discuss cost of maintaining agri-business receivables

# Unit 10

# Compensation Management

## 10.1 Introduction

Compensation is one of the prime needs of all individuals. It is an important motivating factor. Providing fair and equitable salaries and fulfilling employee's needs are essential HR Functions. Compensation management consists of fulfilling government regulations, statutory compliances. There are various methods of regarding employees for their effort. They can be in monetary form or in the form of benefits and services. It plays a significant role in recruiting, retaining and motivating employees. Job evaluation is technique of determining the wages to be paid to employees at different levels. It helps in providing transparent, equitable and consistent rewards to employee.

## 10.2 Meaning

Compensation is an integral part of human resource management which helps in motivating the employees for better performance.

Today compensation structures have come a long way. With changing organisational structures workers need and compensation systems have also been changing. From the bureaucratic Organisations to the participative Organisations, employees have started asking for their rights and appropriate compensations. The higher education standards and higher skills required for the jobs have made the Organisations provide competitive compensations to their employees.

Compensation strategy is derived from the business strategy. The business goals and objectives are aligned with the HR strategies. Then the compensation committee or the concerned authority formulates the compensation strategy. It depends on both internal and external factors as well as the life cycle of an organisation.

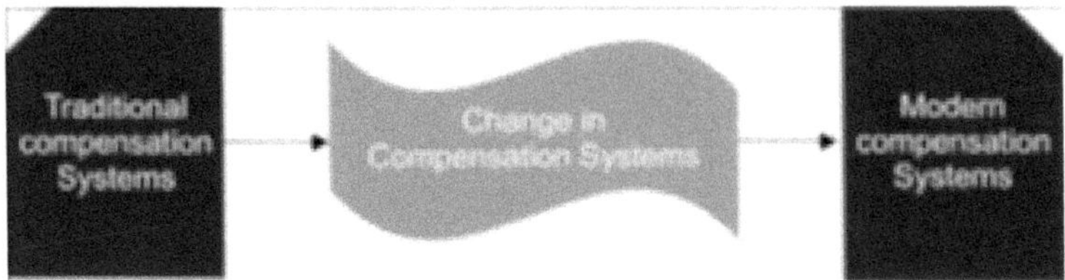


**Fig. 10.1: Evolution of Strategic Compensation**

Their performance was being measured and appraised based on the organisational and individual performance. Competition among employees existed. Employees were expected to work hard to have the job security. The compensation system was designed on the basis of job work and related proficiency of the employee.

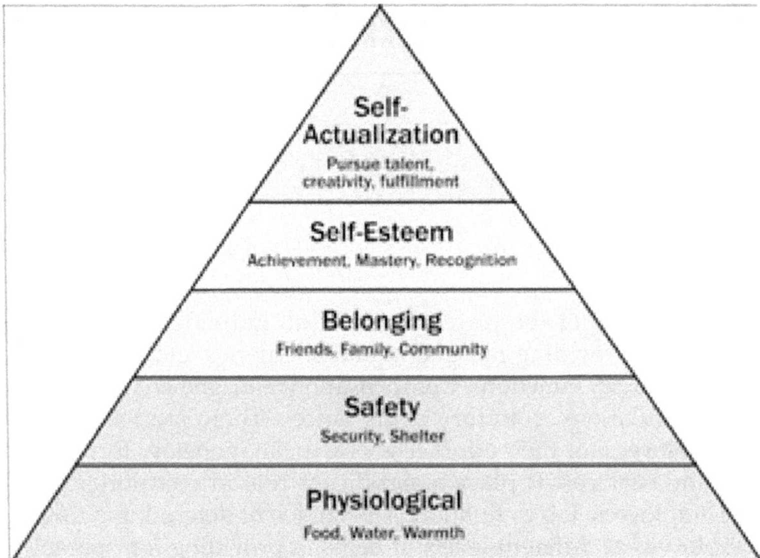


**Fig. 10.2: Maslow's Need Hierarchy**

## 10.3 Today's Modern Compensation Systems

Today the compensation systems are designed in alignment with the business goals and strategies. The employees are expected to work and take their own decisions. Authority is being delegated. Employees feel secure and valued in the organisation. Organisations offer monetary and non-monetary benefits to attract and retain the best talents in the competitive environment. Some of the benefits are special allowances like mobile, company's vehicle; house rent allowances; statutory leaves etc.

### 10.3. 1 Plan

Compensation management is an activity that requires in-depth analysis of various aspects related to manpower, business strategy and organisational vision and mission. There are four main steps to developing a compensation plan:

1. Assessing Total Compensation Implications
2 Plan a comprehensive compensation strategy
3. Implementation
4. Review

#### 1. Assessing a Total Compensation Implications

Total implication refers to the effect of various business aspects on the management of compensation. An organisation s compensation strategy should be determined after analyzing it business strategies and its competitors in the market.

The nature of business plays an essential part in managing the compensation. The concerns like future prospects of the labour market, alignment of strategy with compensation system and comparison of compensation in different countries need to be assessed. For example in case of international operations consideration has to be given to standard working time, number of holidays, country specific benefits etc.

Similarly the cultural aspects are an important consideration as the compensation of a company reflects its ideologies towards people. The behavior of management towards its employees is communicated by the way they are being compensated. The implications can also be assessed related to the:

- ☆ Employees
- ☆ Extent of choice being provided to employees
- ☆ Role of pay in the HR strategy
- ☆ Union management preferences

#### 2. Plan a Comprehensive Compensation Strategy

Designing a comprehensive strategy for compensation is the second step. This consists of various elements of the pay model that mainly consist of four choices:

- ☆ **Alignment:** It refers to the association of pay design with various internal factors like position, experience, hierarchical structures etc. Alignment looks at creating a fine balance between these factors so as to maintain consistency and avoid interpersonal conflicts and confusions.
- ☆ **Performance/output:** Compensation design should motivate employees for higher performance. Employees should be paid on the basis of their contributions to the organisation.
- ☆ **Competitiveness:** Along with managing compensation internally it should also be consistent with the external markets. Pay should be comparable with competitors offering and prevailing wage rates.

### *Role of Management*

Management refers to handling the queries, modes of payments and communication about the pay. HR managers play a crucial role in compensation management as their attitude can create a positive, healthy and competitive environment or can also lead to conflicts. Management of compensations essentially requires answers to the following:

- **Whom to pay:** This requires identifying the target to be paid. It can be an individual employee, a team of employees, department or complete Organizations. It is essential to establish the payment structures for different targets to be paid. Nowadays team work is preferred over individualistic performance. The compensation is different in all these cases and therefore requires proper planning.
- **What to pay:** This is an important consideration and refers to the type of compensation. Whether it has to be just the fixed part on a monthly basis or it can be a mix of variable as well as non-variable compensation like benefits, perks, incentives etc. This requires establishing pay bands and categorizing people based on their skill, knowledge and expertise. What to pay is also important as it is a direct cost to be borne by the company. It also has tax implications.
- **How to rewards:** Different organisation follows different techniques of payments like PBR methods, piece rate system etc. It depends on the company's ideologies and value systems what sort of payment technique they identify with.
- **How to assess:** This refers to evaluation of performance. Based on the nature of jobs companies follow evaluation methods. Some follow MBOs or set up individual KRAs. This part is related with the appraisal of performance and potential of employees

### 3. Implement

The third step is to implement the compensation system in the organisation.

This calls for using reliable communication channels like line and staff managers, notices etc. employees should know the nature of compensation, implication of performance on rewards, modes of payments etc. This step basically requires translating the business strategies and compensation ideologies into practice.

### 4. Review

The last step is to review the implementation for possible loopholes, weaknesses etc. employees feedback and general feasibility of the compensation has to be reviewed so as to avoid any long-term repercussions.

Compensation keeps changing a per market conditions thus reviewing help in realignment and adjustment to the changing conditions.

Compensation management therefore is a continuous activity due to changing employee's needs, market conditions and business strategies. The top management and the HR department plays an important role is establishing the pay structures by boosting employees confidence. This is possible when the compensation structure is:

- **Flexible:** It is able to adjust as per the changes in the market, economy and company internal structures
- **Transparent:** easily understood by employees and self explained
- **Competitive:** So that it motivate employees to perform better.
- Criteria for effective compensation management:
- **Ample:** It should be adequate to fulfill the needs of the employees.
- **Equitable:** It should be consistent with the position, expertise, market rate and skills of the employee.
- **Cost efficient:** It should be consistent with company's budget. Compensation should become a liability to the organisation
- **Motivating:** It should be linked to individual output thereby motivating employees to raise their performance standards.

## 10.4 Components of Compensation System

Compensation system of an organisation is the basis on which employees are paid and rewarded for their contribution. There are many principles on which the compensation system of an organisation is based. Some of them are:

- Create a value for rewards
- Simple to comprehend
- Achievable performance standards
- Participative reward system

An organisations compensation system consists of the following components dimensions:

1. **Pay for work and performance:** This include basic pay, overtime, merit pay b etc. This dimension is related to paying for the fulfillment of the basic responsibility being assigned. Its a monetary form of payment usually on a monthly, weekly or daily base
2. **Pay for time not worked:** Based on the laws employees need to be paid for the time they have not worked. This time can be due to leaves, holidays or weekends. It is an important factor in overall cost to company but is considered worthwhile for making a balance in work life. Usually this constitutes a good number of days due to personal leaves, paid holidays, maternity and paternity leaves etc.
3. **Disability income continuum:** Provisions like sick leaves, ESI, Disability compensation are provided by employers to compensate the families in time of needs. This id due the fact that emergencies can stuck the employee anytime while fulfilling organisation responsibilities and in such conditions when employee is disabled partially or fully his basic needs should be fulfilled.

4. **Deferred Income:** These are income arrangements for employees post retirement. It is provided to employees so that they can manage their life style even after retirement. Deferred incomes are usually in forms of saving schemes, pension fund, annuities, gratuity etc. An important aspect to the deferred income is the tax regulations. They laws related to these incomes are more appealing to the employee. They can defer tax obligation until income tax rates are more favorable. Defer payments also include ESOPs and grant plans that are allowed tax deduction. Thus differed payments are preferred option as it helps employee et more monetary benefits
5. **Health Accidents liability protection:** Companies provide insurance plans to help employees I their medical treatments, payments of bills etc. group insurance plans are opted by employers as they provide better savings.
6. **Income equivalent payments:** Also known as perquisites these provide certain tax benefits both to the employee as well as the employer. Company car, furnished house, child care incentives etc are some form of perquisites given to employee. Perquisites are usually paid on the basis of seniority and experience.

## 10.5 Types of Compensation

Compensation can be provided in many ways. The type of compensation depends on several factors like:

- Type of work
- Budet of the company
- Compensation strategy
- Size and type of company
- Miscellaneous

Broadly Compensation is divided into two types:

1. Direct Compensation
2. Indirect Compensation

### 10.5.1. Direct Compensation

Direct compensation refers to the pay that is given on a regular basis at fixed rate. It is a monetary form of compensation. The various types of direct compensation include:

- ☆ **Base Pay:** It is the stable wage that is provided to employees based on recommendation of the Fair Wage committee. The basic wage is different for different category of workers. It also varies on the basis of location.
- ☆ **Dearness Allowance:** This the monetary allowance provided to employees. It is decided on the basis of market inflations and cost of living index. It is a means to protect the wage earners and help them sustain their living.
- ☆ **House Rent Allowance:** Allowances are fixed sum of rupees given to employees for special needs like house rent, medical etc. The HRA is fixed

on the basis of government regulations. However it is not mandatory for employers to provide it. Government employees et this at a fixed rate.

- **Conveyance:** This is an allowance provided to employees for traveling for office purpose. It's a taxable pay and is exempted upto Rs. 800.
- **Bonus:** it is a sum provided by employers for sharing the profits of a company. It is an important pay component for the industrial workers. Payment of bonus is mandatory under the payment of bonus act 1965.
- **Leave Travel Allowance:** Employees are encouraged to go on compulsory leaves with family for which the company provided reimbursements. The payments are usually on yearly basis or once in two years. Depending on the company HR policy.
- **Gratuity:** It's a lump sum provided to employees who have completed five years of continuous services with the company. The gratuity is calculated on the basis of total completed years of service and the last salary drawn (basic and DA). It is paid at the time o f retirement or while leaving the company.

**Fig. 10.3: Direct Compensation**

## 10.5.2. Indirect Compensation

Indirect compensation is usually in the form benefits or services provided by the employer. The indirect compensation is provided to employees based on their grades, performance and levels. This type of compensation is provided to fulfill a government regulation or to motivate employees. Flexible timings:

- **Insurance:** Employees are covered under insurance policies to take care of their health issues. Insurance may be provided for medical as well as life benefits. The company may bear the whole amount of insurance or it can be paid in equal proportion by the employer as well as the employee. Usually field staff, and people who are prone to risk due to extensive travelling are provided this type of benefits.

- **Transportation:** Companies provide transportation facilities to outs station employees.
- **Provident Fund:** It's a statutory compliance to be fulfilled by all the employers who are covered as per the eligibility. It is managed by the central government. PF is a saving scheme for retirement that is paid with equal contribution of the employer and the worker. The rate of POF is 8% on basic and DA.
- **Employees State Insurance scheme (ESI):** This is a health insurance benefit mainly for the industrial workers run by the central government. Employee and their dependents are provided free medical services incase of maternity, sickness and death.
- **Crèches:** For working mothers who have small kids companies provide crèche facility where employee can bring their children. They are provided time to visit and take are of the child needs.

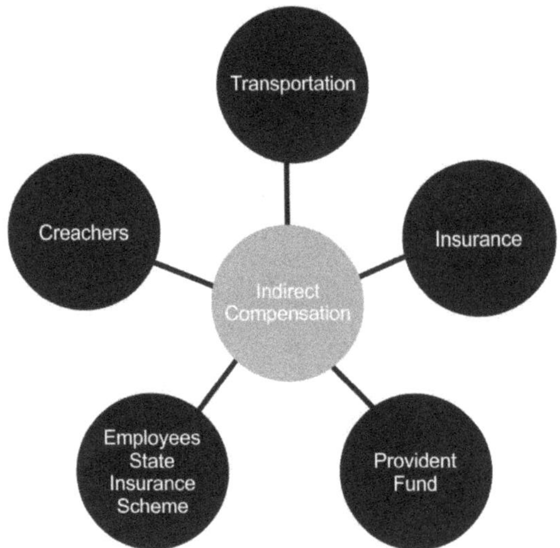


**Fig. 10.4: Indirect Compensation**

- Relational Returns
- Recognition
- Promotion Opportunities
- Working Conditions
- Interesting Work

## 10.6 Compensation Process

The process of compensation aims at aligning the rewards, pay and incentives with organisational goals. It's a strategic process that comprises of in depth planning.

The process comprises of the following steps:

1. **Analyzing business strategy -** The compensation structure of an organisation strongly depends on its business structures, plans and objectives. The compensation varies depending upon the size of business. A mid size company will provide lower salaries as compared to that of big conglomerates. Similarly if a company plans to expand its operations it will go for higher and competitive compensation and rewards as their focus will be on acquiring experienced people. They also need to retain them. The industry to which a company belongs also affects its compensation policy.

2. **Organizational policy -** Every company is driven by their specific sets of values, philosophies and culture. These have a high impact on its compensation policy. An organisation similarly also have to adhere to the external and internal expectation of markets, government and employees. An organisational policy comprises of the following:

   (a) HR Policy refers to standardized employee practices followed by a company's HR department. The HR policy for compensation usually determines the pay structure based on person, performance and position. These are referred as 3Ps of compensation.

   (b) Compensation policy of an organisation is decided with due consideration to the external & Internal environment. External environment includes government policies, market rates, labour relation etc. Internal condition refers to the various types of jobs, departments, and people employed.

**Fig. 10.5: Compensation management process**

3. **Job evaluation -** The next step is to evaluate the various jobs in an organisation. Job evaluation will be discussed at length in the next topic. It helps in ascertaining the relative worth of jobs. Based on the results jobs are categorized and labeled under different pay grades. Pay grades are determined by various methods.

4. **Designing compensation plan -** After the value of different jobs has been determined the next important step comprises of developing a comprehensive compensation plan. Compensation plan includes the following:
   (a) Salary Grades/ Broad banding.
   (b) Pay structures is reflects the overall design of salary consisting of various salary components.
   (c) Components of pay structure. Like Basic salary, incentives, benefits, taxes, allowances etc.
5. **Implementation and review -** Once the Compensation plan is developed it is formally communicated to all the line managers for sharing it with employees. The compensation plan is implemented and reviewed periodically to find out any loopholes.

## 10.7 Techniques of Job Evaluation

Job evaluation determines the worth of the jobs in a company relative to one another. Evaluation analyzes the value of jobs and is the source of targeted job descriptions. Job evaluation assesses the differences in contributions between jobs, job content and value and provides the basis for compensation decisions. Job evaluations are done by human resources departments, compensation specialists, consultants, auditors, managers and supervisors. They are part of a compensation system and can be used as the basis not only for pay but also to determine comparable worth, to create job descriptions, to develop titles, to plan for employee training, development and succession and for business organisation. Job evaluation is a valuable analytical tool to manage a business. There are mainly four ways of evaluating jobs. These are divided on the basis of:

### 10.7.1 Non-Quantitative Method

In these methods a job is considered as a whole while comparing with other jobs.

#### Ranking Method

The ranking method is simplest of all the methods. It consists of comparing two jobs as a whole for example job of machine operator will be compared to job of supervisor as a whole. The main criteria if comparison is usually the role and the responsibilities of that job. All the jobs are ranked in order of their importance.

This method is useful in small size companies as there are fewer jobs that can easily be compared. Though this method is easy and simple, its biggest demerit is the absence of standard system of evaluation that can lead to biases.

#### Classification Method

This method involves diving various jobs in different categories based on the nature, role and responsibilities. Jobs are classified in different grades. Job grades are a common term used in compensation and reward management and refer to group of jobs with similar skills, behaviors, and knowledge. For example, computer operators, clerks, admin executive can be kept in one grade as they nature of work is almost similar.

The grades are developed by experts who study the jobs thoroughly. Once the job grades are developed, the next step is to assign jobs to their appropriate grades. This method helps in compensation management as well as tracking the performance. However this method is not suitable incase of higher level and complex jobs.

## 10.7.2 Quantitative Methods

In this method, the key factors of a job are selected for measurement.

### Point Method

This is the most widely used method of job evolution. In this method jobs are broken into identifiable factors like skills, efforts, training, knowledge etc. after this, Points are located to each of these factors. Weights are given top factors depending on their importance to perform the job. Points so allocated to various jobs are added. The jobs havening similar total points are placed in one grade pay.

The sum of the points gives n index of the relative significance of the jobs that are rated. This is the most sophisticated method of job evaluation and based on facts and Figures. There is no possibility of the results getting challenged. The procedure involves determining job points as follows:

1. Determine the jobs to be evaluated usually covering all major occupations and levels of responsibility
2. Decide on the factors to be used in analysing and evaluating the jobs. The number of factors to restricted to avoid duplication
3. Define and write down the factors so that raters can clearly understand them
4. Determine degrees of each factor and assign point value
5. Point values are assigned to different degrees on the basis of arithmetic progression
6. Last step is to assign money value to pints. Points are added to provide total value of jobs and translated into money terms.

### Factor Comparison

This method is a combination of Ranking method and point rating method. It compresses the two jobs as well as analyse them by breaking the jobs into compensable factors. Further, base rate are determined and allocated to common factors of the jobs in question. This method is objective in nature and help in determining the relative worth of a job. It is mainly used for rating executive level and professional's jobs.

There are many more unconventional methods that being used now days like, Urwick Orr profile method, Decision Band method, the direct consensus methods, and any more such methods. These modern methods use decision making, problem solving, and accountability as factors for comparing and rating the jobs.

### Motivation and Fringe Benefits

Every person has different motivations for working. The reasons for working are as individual as the person himself. But, we all work because we get something in return. That something we obtain from work impacts our morale and motivation and the quality of our lives. Here is the most recent thinking about motivation, what people want from work.

Some people work for love; others work for personal fulfillment. Others like to accomplish goals and feel as if they are contributing to something larger than themselves, something important. Some people have personal missions they accomplish through meaningful work. Others truly love what they do or the clients they serve. Some like the camaraderie and interaction with customers and coworkers. Other people like to fill their time with activity. Some workers like change, challenge and diverse problems to solve. Motivation is individual and diverse. Whatever our personal reasons for working, the bottom line, however, is that almost everyone works for money. Whatever we call it: compensation, salary, bonuses, benefits or remuneration, money pays the bills. Money provides housing, gives children clothing and food, sends teens to college and allows leisure activities and eventually, retirement. To underplay the importance of money and benefits as motivation for people who work is a mistake.

One of the important parts of compensation is the Fringe benefiters. These are extra benefits in monetary or non monetary for provided to employees irrespective of their output and performance. These are kind of facilities provided to employees to motivate them. Fringes are in addition to the employee basic compensation and are an additional cost to the company.

As per the American Heritage Dictionary "fringe benefits are the additional benefits given in addition to the wages and salary". Weather & Davis defines fringe benefits as "These embrace a range of benefits and services that employee receive as part of their total compensation package. Benefits and services however are indirect compensation because they are usually extended as a condition of employment and not directly related to performance". Generally fringe benefits are either statutory, that has to be compulsorily provided as per law or voluntary that the organisation provides at their own will.

Following are some forms of fringe benefits given by employers:

- Lon service awards in cash and kind
- Employment opportunities for children of employees
- Thrift societies for payment of loans
- Housing schemes at lower rates
- Educational assistance, like fees sponsorship to meritorious students
- Transport facilities like pick and drop, cab etc
- Food vouchers at office canteens
- Crèches for children
- Medical and health centers in the company's campus
- Health club membership

The above list of fringe benefits displays an array of employee needs related to economic, social, family and future security. Motivation as we all know is a drive that instigates an individual to work. Various theories of motivation indicate the direct relation of performance with the fulfillment of these needs.

The compensation structures therefore try to meets these needs of employees by providing them Fringe benefits that develops commitment towards employers. Fringes are of different types and for different levels of employees. For example ESOPs, Profit sharing may for those employees who are directly linked to increasing company's revenue. The benefits for executive, managers and shop workers may vary accordingly.

# Questions

## Short Answers

1. What is Compensation strategy?
2. Idea about External equity.
3. What are the components of compensation?
4. Gratuity, Provident fund are example of
   (a) Health Protection c. Employment protection
   (b) Old Age Protection d. None of these
5. In case of ____ employees are entitled to compensation at the rate to 50% of the total of the basic wage and dearness allowance.
   (a) Retrenchment c. Closure
   (b) Retirement d. Lay off

## Long Answers

1. Explain types of compensation in agri-business?
2. Features of HR interventions in agri-business.

# Glossary

**Broadband :** It is a procedure of putting jobs of similar nature and capabilities across different verticals in one group. It is advancement to salary grades and aims at reducing hierarchical levels in the organisation.

**Crèches :** These are kind of nurseries where children can stay for short times. Companies provide this facility to their female employee who has small children who cannot be left at home.

**Salary Survey :** It is a study of the current market rates for wages of different categories of employees. It helps in determining equitable salaries for employee.

**Fringe benefits :** This refers to the additional advantage being given to employees other than their regular income. It usually includes perks like company car, paid holidays, subsidized meals, education assistance. These are usually associated with the level of job and may be taxable.

**Retrenchment :** It is the technique of reducing the excess workforce from the organisation. There are some legal implications like retrenchment has to be cases other than that of punishment. It is a strategic decision taken by the management.

## Unit 11

# HRM Trainings

### 11.1 Human Resources

Human resources is the job function that manages people in an organization by providing an assortment of activities and policies and procedures, which relate to developing, utilizing, maintaining and retaining the appropriate number, and skills, of employees to accomplish your businesses objectives and goals. You can implement human resources into your agri business by creating a human resources plan. It can include:

- ✰ Employee recruitment and selection
- ✰ Training and development
- ✰ Organizational structure (who reports to whom/pecking order, job skills and knowledge)
- ✰ Labour relations (managing compliance with various legislation and regulations)
- ✰ Employee performance management and succession planning
- ✰ Human relations (discipline, performance management, complaint handling, counselling and coaching)
- ✰ Employee benefits
- ✰ Health and safety
- ✰ Employee communications

- Strategic planning
- Management of employee records

### 11.1.1 Human Resource Evaluation

A HR evaluation is an assessment of HR policies, processes, documentation and procedures of a agri business that helps ensure the HR plan is working efficiently, and identifies areas that need improvement. An HR evaluation can help assess:

- The hiring process
- Employee retention
- Training
- Employee compensation
- Management and employee relations
- Processes or practices that affect an organization's people

This is to account for your workforce, and the efficiency with which your agri-business deals with its people – from the beginning to the end of their employment with you. This process allows a agri business to get a general idea of where its HR currently stands, where it can be corrected or improved, and how to plan for the future. It also helps prevent employee issues that may directly impact your profitability. Managing your workforce gives your business a competitive advantage, especially when labour and skills are limited. Assessing the effectiveness of your current HR practices can help ensure that you attract and retain not just qualified workers, but the best people for the job.

## 11.2 Developing an Identity for Your Agri Business

A crucial piece of your agri-business's identity is the message (description) you share with your suppliers, employees, community, industry, and others. In marketing and advertising, this message is known as your business's boilerplate or profile. Those values determine the way we work, the quality we offer, and the unsurpassed treatment you get as a customer, investor, and employee.

Your agri-business's profile should be about a paragraph long (three to four sentences) and should combine:

- Useful information (ex: product information, year founded, location)
- Your agri-business's personality (ex: values, beliefs)
- Anything that is unique about your agri-business

When creating your agri-business's profile, think of:

- Your agri-business's qualities or the values that are important to you
- The way other people think about your agri-business
- The way you think about your agri-business
- The emotions and perceptions that you elicit, or want to elicit, in other people.

A strong agri-business profile or description will help by allowing candidates to recognize why they want to work for you as opposed to the agri-business down the road. Not only does this profile yield external benefits – it also helps you internally by aligning your identity with your employees and the business decisions you make.

## 11.3 Recruiting

Recruitment is the process of planning, selecting, and hiring employees. The goal is to identify and hire the most suitable employees you can find. The process starts with a marketable agri-business identity and a pre-planned recruitment process and strategy.

Recruiting means becoming involved in finding and attracting employees. Finding includes seeking, sourcing and locating employees. Many agri-business struggle with recruiting – often believing that they can't find good people. Recruiting employees requires a strategy, effort and commitment. Think strategically about:

- Skills needed in your agri-business business.
- The type of person and behaviours that fit best with your leadership style and the culture of your agri business.
- What value your agri business offers a successful candidate.
- Key places to advertise so you can find the right person.

Recruiting is a proactive process versus a reactive process. In fact, many agri businesses that hire reactively (*i.e.* when they are in a panic for labour during harvest), end up with bad hires that ultimately cost the business money. By hiring proactively, you'll understand what job needs to be filled, the skill sets required, the personality attributes important for the position, and what type of person would best fit with you and your team.

Recruiting has changed in the past few years, particularly in the agri-business industry. As agri-business continue to grow, and some rural populations decline, it may be harder to find staff nearby with the right skills for the job. Many of the traditional skills have become more sophisticated. Also, today's younger generation wants to find employment with a business that offers more than a pay cheque. Use your brand statement, and the qualities that differentiate you from other agri-business, to create job ads that are exciting, interesting and appealing. You need to let potential employees know not only what you're looking for, but what you can offer them.

## 11.4 Hiring

There are few things in business that are more important than hiring the right people. Some say that without the right people, no amount of money can make a business succeed. Likewise, your agri-business depends on qualified staff to make it run smoothly.

Traditionally, agri-business simply hired someone that another agri-business owner recommended. It was quick and easy – no need to go through applications or interviews. However, it was also common for the individual hired to have had little or no experience, which led to trouble later on.

To create a hiring process, follow these four simple steps:

1. Establish who on your agri-business should be involved in shortlisting and interviewing candidates.
2. Review the applications.
3. Shortlist applicants.
4. Interview

## 11.4.1 Steps

### Step 1

Establish who from your agri-business should be involved in shortlisting and interviewing candidates As the owner, you may not always be involved in the day-to-day activities of each job. Determine who will be, and include these people in shortlisting and interviewing the candidates. If staff are involved in hiring, it can make them more accountable for the new hire's success.

For example:

- ✰ If the position is for a general labourer, perhaps the agri-business owner and position's supervisor need to be involved.
- ✰ If the position is for a agri-business manager, perhaps the agri-business owner and family need to be involved.

Make sure each person involved in any step of the hiring process:

- ✰ Has a copy of the job description
- ✰ Understands the position's role and accountabilities
- ✰ Understands the selection criteria

Also, ensure that you designate associated tasks to a staff member, including receiving resumes, reviewing the applications, shortlisting, pre-screening and interviewing.

### Step 2: Review the Applications

You can start the review process in two ways:

1. You can have a formal discussion with staff who are directly involved in hiring for the position.
2. You can distribute a folder, containing copies of the applications and resumes, to each person involved in the selection process.

We recommend that you review all of the applications at the same time, so you can compare them. Also, have a list of what the job requires to ensure an applicant fits the position. It's easy to get excited about an applicant who has hobbies and characteristics similar to yours. However, you need to ensure he or she also has the skills to fill the position.

**Step 3: Shortlist the Candidates**

After reviewing the feedback – either as a group or with the person in charge of hiring – you can shortlist the candidates you wish to interview. Select two to three of the best candidates for each position.

**Step 4: The Interview Process**

Now that you've shortlisted the candidates, you can start the interview process.

## 11.5 Employee Orientation and Training

Orientation and training helps you incorporate new employees into the culture of your agri-business, so they can become more productive on your agri-business and in their jobs. It also creates a faster employee ROI (return on investment), also known as return on individual. The time you put in now to orient and train your employee will be returned in how quickly the employee can become a productive worker.

Orientation and training also helps your new employees work smarter and safer. When an employee can't work due to an accident, it can cost you money, time, and stress. When employees are trained to perform their jobs in the correct manner from the beginning of their employment, they will have less frustration, better morale, higher productivity and safety.

New employees will probably have a desire to succeed and be anxious about working in a new environment. A strong new employee orientation and training program will show them they've made the right decision to work on your agri-business.

### 11.5.1 Orientation and Training Tips

1. Use a new employee orientation checklist to ensure you don't forget anything (we've included an example in the Forms Appendix – it's important to customize this checklist for your agri-business).
2. Explain how and why a particular task is performed.
3. Demonstrate the correct way of performing a task. Review the task at a normal pace and repeat it at a slower pace, pointing out the various steps along the way, and answering questions.
4. Point out potential hazards associated with the job and ensure you explain safety procedures and regulations. Check to make sure the new employee understands them.
5. Have the new employee perform the task while you watch.
6. Check in with the new employee often for the first few days or weeks.
7. Consider having someone interpret your orientation and training if you hire employees who may not speak English as a first language.

### 11.5.2 Creating a Training Plan

Although you may hire employees who are already experienced in the position, they will still need training to do the tasks to your standards and expectations.

When creating the training plan for your new employee orientation and training program, you should consider:

1. The specific tasks for the position
2. Whether the tasks are better taught with an explanation or a demonstration
3. The time required to teach each task
4. What result the employee is trying to achieve by performing the task, what impact that task has on the agri-business business and other positions, and why it's important
5. Incorporating a measurement to ensure the employee understands and can perform the task (for instance, you may want the new employee to perform the task while you watch)

## 11.6 Communication

Communication is the sharing of ideas and information. It's an essential part of work and life and, at times, it's not easy to do. Differences in people's personalities, communication styles and skills can play a part in how well we communicate with each other. Being an effective communicator starts with an understanding of how you communicate. Below is a list of communication skills to think about and work on:

1. **Message:** Know what message you want to communicate. Organize your thoughts so that your message will be clear and easy to understand. Unorganized thoughts can lead to misunderstandings and be confusing to the listener.
2. **Plan:** Important conversations should be planned. When planning a conversation, think about the different scenarios, reactions and outcomes that can occur and also the personality and behaviour of the person you will be communicating with. Plan what you will say and do with each reaction, and prepare a solution for each reaction, so that you know beforehand how you will respond.
3. **Body Language:** Non-verbal signals (body language) play a significant role in communication. Your facial expression, posture and gestures directly impact your message. Your body language should be aligned with your words in order for your communication to be clear.
4. **Positive language and tone:** Stay positive during the conversation. Negative statements often elicit a negative reaction, while positive statements often elicit a positive response.
5. **Listening:** It's very important that you listen to what the other person has to say.

### 11.6.1 Tools Used

#### Bulletin Board

- ☆ Create a bulletin board to hang in your main office, barn, etc. Make sure it is regularly

- Updated with information about the agri-business, employee news and any new or updated processes and procedures.
- To organize the board, you can section it into four sections (quadrants) and have each
- Section represent a different topic.
- If you're technologically savvy, you can also create a bulletin board online, using a simple blog interface, like Blogger or WordPress.

**Memos**

- A memo is still an effective way to communicate with staff – especially on a agri-business where email may not be commonplace.
- Prepare a weekly, bi-weekly or monthly memo that is distributed to all agri-business employees and family.
- You can also include the memo with your employee's paycheques.

**Agri-business Post Office**

- Setup a post office on your agri-business.
- Find an area to set up mailboxes or slots – one for each employee.
- Ensure that you set it up somewhere your employees or family visit daily.
- You can update your workers about agri-business matters by delivering information to their mailboxes.
- As part of this process, you might want to set up a mailbox where staff can leave you
- Anonymous comments (like a suggestion box).

**Email / Text Messages**

- Not every business has the luxury of mobile email or text messaging, but if you do, you can update and communicate with employees through these devices.
- Remember to always give the email or text message a consistent subject line (ex: agri-business updates), so employees know it's a communication coming from you about the agri-business. Also consider the time of day when you send the message. Generally, employees will appreciate
- Getting text messages more during working hours than late at night.

## 11.7 Motivate Employee Performance

An employee needs motivation. Motivation is simply the willingness to achieve a goal – to get something done. Motivation is not a personality trait as we are all motivated by different things. Understanding what motivates an employee is difficult – for one employee, it might be work-life balance and, for another, it may be money. Some characteristics of motivated people are those who:

- ☆ Enjoy their work
- ☆ Work well in a team and co-operate to get things done
- ☆ Focus on achieving results
- ☆ Never say: "That's not my job" or "I can't help you"
- ☆ Take an interest in their surroundings (ex: take care of and clean their equipment, workspace, etc.)
- ☆ Celebrate successes and don't blame others when something goes wrong
- ☆ Are reliable, punctual and attentive
- ☆ Ask questions and want to learn more

There are some common motivators and demotivators that you should be aware of when trying to motivate an employee. Demotivators directly impact performance and are often minor, daily activities that frustrate employees and affect their performance, consciously or unconsciously. Demotivators weaken morale and affect almost everyone, as well as the operation's profit line, over the long term.

### 11.7.1 Performance Review Tips

#### *Meeting day*

Book the performance review meeting with the employee. Make sure you give the employee two to three weeks' notice before the meeting. Ensure that it's held in a private room where you and the employee won't be interrupted.

#### *Set the Tone*

Start the discussion with a friendly greeting and upbeat attitude. This will set the mood for the rest of the meeting.

#### *Outline the Meeting*

Let employees know what the meeting is about and the areas you will cover.

#### *Focus on Performance*

Keep the conversation and feedback on your employee's performance, in terms of finishing tasks, achieving results, and handling work situations.

#### *Feedback and Discussion*

Go through each section of the performance review form (sample on next page) with your employee and provide feedback and specific information about why he or she received the rating. Also, make sure you ask the employee if he or she has any questions. Encourage feedback, and write it down.

#### *Listen Actively*

Make sure you understand what your employees say by rephrasing, summarizing, and writing down their comments.

***End the Meeting***

Summarize the discussion, ask the employee if he or she has any final questions, set follow dates for setting goals, and make sure employees sign a hard copy of the performance evaluation. Make sure to provide the employee with a copy and keep one for your files.

## 11.8 Employee Manuals

An employee manual and SOPs manual gives people working on your agri-business a sense of value, structure, and allows them to understand the policies and procedures of your agri-business, regardless if they are family or and outside employee. It sets the tone and standards about "how things are done" within your agri-business. Whether they are family or non-family, they must follow the same policies and procedures.

There are a number of reasons why you should have an employee manual, including:

- Having clear uniformity and standards
- Saving your managers time with a well-planned and written manual, so they don't have to re-explain policies
- Giving employees a written resource that clearly explains information about your policies, including holidays, guidelines, leave, work hours, overtime, pay procedures, employee safety measures and procedures

## 11.9 Resolving Employee Conflict

As a manager, you must deal with conflicts, difficult employees and workplace issues when they arise. A good manager will have a plan in advance to ensure problems don't impact the agri-business. Ignoring these issues can have dire consequences on employees, their performance and the business. Overlooking conflict will usually:

- Allow it to grow, get more complicated and be harder to resolve
- Negatively impact morale
- Reduce employees' confidence and trust in you as a manger
- Increase turnover and affect the reputation of your agri-business
- Negatively affect employee performance and productivity

It's important to remember that some employee conflict may never truly be resolved. Employers should be encouraging employees to work together in a professional manner.

To resolve this conflict, I've determined:

- Who is involved in the conflict
- Why this particular conflict is occurring
- A policy in the employee handbook that addresses this conflict
- Whether I want to intervene in the conflict or let the employees work things out on their own

- ☆ Consequences of intervening and not intervening
- ☆ Whether I need to consult with an outside party for advice (ex: another agri-business owner whose judgment you trust, an outside conflict management contractor)
- ☆ Appropriate conflict resolution that takes into account the dignity and rights of individuals

## 11.10 Handling Discipline Issues Effectively

Dealing with performance problems and taking disciplinary action can be frustrating and is one of the stressful issues that management faces. You should have an established and consistent process for dealing with discipline. Your goal is to work out a strategy so that your employee becomes an effective member of your team.

An employee may have legitimate reasons for the issue (ex: medical, personal). Don't jump to conclusions until you sit down and gather the facts. This will also help you build a relationship with the employee.

Below is how a typical progressive disciplinary process is structured:

1. **Initial Notification:** This is the first step in a progressive warning. It informs the employee that his/her job performance or work conduct isn't measuring up to your operation's standards. The owner, or the person's manager, should deliver this initial warning at a one-on-one meeting. At that time, you should complete the first Official Disciplinary Notice.
2. **Second Warning:** This applies if the performance or conduct continues or worsens. The owner or person's manager should hold another one-on-one meeting to discuss the performance issue and complete the second Official Disciplinary Notice. Make sure you let the employee know his/her performance is affecting the business and ensure that he/she understands what is being communicated. You should also create a written action plan with the employee, which provides concrete goals and a timeline for achieving them.
3. **Final Warning:** This warning informs the employee that if the performance or conduct does not improve, the employee will be subject to termination.
4. **Termination:** This is the last step in the process when all other corrective or disciplinary actions have failed to resolve the problem. The tone of a termination meeting should be one of cordiality and empathy. In some cases, the best way to start the meeting is to say something like, "You will probably not be surprised to find out that things are just not working out."

# Glossary

**Antenna Recruiting :** Observe people around you by watching their attitudes, skills and behaviours when you see people who suit your agri-business, remember who they are - they could possibly be employees now or in the future.

**Background Checks :** Background checks objectively evaluate a job candidate's qualifications, Character and fitness, and identify any potential safety and security hiring risks.

**Behavioural-based Interview :** A job interview focused on discovering how an applicant acted in Specific employment-related situations. Instead of asking "How would you behave?" the interviewer will ask "How did you behave?" The interviewer wants to know how a candidate handled a certain situation, instead of what the candidate might do in a certain situation.

**Brand :** It's the emotional and psychological relationship you have with your customers and the Personality of your agri-business. Strong brands elicit thoughts, emotions and, sometimes, physiological responses from customers.

**Candidate :** A person who applies for a job.

**Candidate Testing :** Testing candidates before you hire them will help you accurately assess their skills, training needs and suitability. There are several companies that offer pre-made tests for all sorts of positions in a variety of industries.

**Compensation and Benefits :** The total amount of money and benefits (ex: dental insurance, vehicle allowance) provided to an employee by an employer in return for work performed as required.

**Core Competencies :** The behaviours, skills and knowledge a person is expected to demonstrate and perform to fulfill a job position.

**Culture :** The values and practices shared by members of a group.

**Compensation and Benefits :** The total amount of money and benefits (ex: dental insurance, vehicle allowance) provided to an employee by an employer in return for work performed as required.

**Core Competencies :** The behaviours, skills and knowledge a person is expected to demonstrate and perform to fulfill a job position.

**Culture :** The values and practices shared by members of a group.

**Employer of Choice :** A place where people want to work and remain working for many years (because the employee enjoys the workplace and chooses to work there).

**Employee Referral Bonus :** A system where existing employees recommend prospective candidates for jobs offered in an organization. If the suggested candidate is hired, the employee who referred the candidate receives a bonus. A bonus can be in the form of cash, gift certificates or other incentives.

**New Immigrant Considerations :** Immigration in Canada continues to grow. There are several Internationally trained individuals who can provide your agri-business with a competitive edge in new global and ethno-cultural markets domestically. There are also several tax incentives for hiring a new immigrant to Canada.

**New Employee Orientation :** An on-the-job introduction for new employees to gain the necessary Knowledge, skills, and behaviours they need to become effective employees. This might include a period of mentorship, an orientation session, a tour and/or a recap of benefits and policies.

**Psychological Testing :** Written, visual, or verbal evaluations given to assess the cognitive and emotional functioning of a person. They are used to assess a variety of mental abilities and attributes, including achievement and ability, personality and neurological functioning.

**Social Media :** It's an online platform where you can build an audience to speak on topics, and spread your information.

**Standard Operating Procedures (SOPs) :** A written document or instruction that details all the steps and activities required to complete a process or procedure.

**Structured Interview :** This type of interview allows you to collect responses from each candidate you interview and compare responses to hire the best person for the job.

# References


Albrecht, K. 1980. Brain Power: Learning to Improve Your Thinking Skills. New York: Simon and Schuster.

Allaire, Y., and M. E. Firsirotu, M.E. 1984. Theories of organizational culture. *Organization Studies* 5:193-226.

Allen, R.W., et al. 1979. Organizational politics:tactics and characteristics of its actors. *California Management* Review 22: 77-83.

Andrews, Kenneth. 1989. Ethics in practice. *Harvard Business Review* (Sept- Oct): 99-104.

Argyris, Chris. 1987. Double loop learning in organizations. *Harvard Business Review (Sept-Oct)*: 115-125.

Ashforth, B.E., and F. Mael. Social identity theory and the organization. *Academy of Management Review* 14 (1): 20-39.

Ashmos, D.P., and G.P Huber. 1987 The systems paradigm in organizational theory: Correcting the record and suggesting the future. *Academy of Management Review* 12 (4): 607-621.

Augustine, Norman R. 1987. Reshaping an industry: lockheed martin's survival story. *Harvard Business Review* (May-June).

Barasch, Douglas, S. 1987. God and toothpaste. New York Times Magazine.

Barrick, M. R., and M.K. Mount. 1991. The big five personality dimensions and job performance: a meta-analysis. *Personnel Psychology* 44: 1-26.

Bass, B. M. 1985. Leadership and performance beyond expectation. New York: Free Press.

Becker, H.S., and B. Geer. Latent culture. *Administrative Science Quarterly* 5: 303-313.

Bedke, Curtis M. 1993. Strategic decisionmaking in a multinational ad hoc coalition ad astra per aspera. Unpublished ICAF course paper, 17 December: 9.

Bennis, Warren. 1989. On Becoming a Leader. Reading, MA: Addison-Wesley Publishing Co., Inc.

Bennis, W. and Nannus, B. 1985. Leaders: The Strategies for Taking Charge. New York: Harper and Row.

Bhide, Amar, and Howard H. Stevenson. 1990. Why be honest if honesty doesn't pay? *Harvard Business Review* (Sept-Oct): 21-129.

Blau, P.M. 1964. Exhange and Power in Social Life. New York: John Wiley.

Bolman, L. G., and T.E. Deal. 1991. Reframing Organizations: Artistry, Choice, and Leadership. San Francisco, CA: Jossey-Bass Publishers.

Bourgeois, L.J. III, and Kathleen M. Eisenhardt. 1987. Strategic decision processes in silicon valley: the anatomy of the living dead. *California Management Review* (Fall): 143-145.

Bourgeois, L. J. III. 1985. Strategic management and determinism. *Academy of Management Review* 9: 586-596.

Briggs Myers, I., and M. McCauley. 1985. Manual: A Guide to the Development and Use of the Myers-Briggs Type Indicator. Palo Alto, CA: Consulting Psychologists Press, Inc.

Briggs Myers, I. 1992. Introduction to Type. Palo Alto, CA: Consulting Psychologists Press, Inc.

Brilhart, John K., and Gloria J. Galanes. 1989. Effective Group Decisions Dubuque, IA: William C Brown Publishers: 201-203.

Broadbent, D. E. 1977. Levels, hierarchies, and the locus of control. *Quarterly Journal of Experimental Psychology* 29: 181-201.

Brown, L.D. 1983 Managing Conflict at Organizational Interfaces. Reading: Addison-Wesley.

Brusilov, A. A. 1931. A Soldier's Notebook: 1914-1918. Westport: Greenwood Press, Publishers.

Builder, C.H. The Masks of War: American Military Styles in Strategy and Analysis. Baltimore: The John Hopkins University Press, 1989.

Caminiti, S. 1995. What team leaders need to know. *Fortune*, Feb 20: 94, 100.

Chaloupka, Mel G. 1987. Ethical responses: how to influence one's organization. Naval War College Review (Winter): 80-90.

Chandler, Clay. 1997. Washingt Post, March 28: A 27.

Cherrington, J. O., and D. J. Cherrington. 1992. "A menu of moral issues: one week in the life of the Wall Street Journal. *Journal of Business Ethics* II 4: 255-265.

Church, A. H. 1997. Managerial self-awareness in high performing individuals in organizations. *Journal of Applied Psychology* 82 (2): 281-292.

Clark, Kim B. and Wheelwright, Steven C. 1992. Organizaing and leading 'heavyweight' development teams. *California Management Review (Spring)*: 9-28.

Collins, James C. and Jerry I. Porras. 1991. Organizational vision and visionary organizations. *California Management Review* (Fall): 30-52.

Colosi, Thomas R. 1993. On and off the record: Colosi on negotiation. Dubuke, IO: Kendall/Hunt Publishing Company.

Conger, J. A. 1989. The Charismatic Leader: Behind the Mystique of Exceptional Leadership. San Francisco, CA: Jossey-Bass Publishers.

Cosier, Richard A., and Charles R. Schwenk. 1990. Agreement and thinking alike: ingredients to poor decisions. *Academy of Management Executive* 4 (1): 69-74.

Costa, P. T., and R.R. McCrae.1992. Trait theory comes of age. In Vol. 30, Psychology and Aging: Current Theory and Research in Motivation. Edited by T. D. Sonderegger. Lincoln, NE: University of Nebraska Press.

Daft, R.L., J. Sormunen, J. and D. Parks. 1988. Chief executive scanning, environmental characteristics, and company performance: An empirical study. *Strategic Management Journal* 9: 123-139.

Daft, R.L., and K.E. Weick. 1984. Toward a model of organizations as interpretation systems. *Academy of Management Review* 9: 284-295.

Dahl, R.A. 1957. The Concept of power. Behavioral Science 2: 201-215.

Dess, Gregory G. 1987. Consensus on strategy formulation and organizational performance: competitors in a fragmented industry. *Strategic Management Journal* 8: 259-277.

Drucker, P.E. 1974. Management: Tasks, Responsibilities, Practices. New York: Harper and Row : 466-467.

Dunegan, Kenneth J. Framing. Cognitive modes, and image theory: toward an understanding of a glass half full. *Journal of Applied Psychology* 78 (3): 491-503.

Eisenhardt, Kathleen M. 1989. Making fast strategic decisions in high-velocity environments. *Academy of Management Journa*l 32 (3): 543-576.

Eisenhardt, Kathleen M., and Bourgeois III. 1988. Politics of strategic deicdsion making in high-velocity environments: toward a midrange theory. *Academy of Management Journal* 31 (4): 742-753.

Eisenhardt, Kathleen M., and Mark J. Zbarecki. 1992. Strategic decision making. *Strategic Management Journal* 13: 20-22.

Eisenhardt, Kathleen M. 1990. Speed and strategic choice: how managers accelerate decision making. California Management Review (Spring): 39-54.

Fairhurst, Gail T., and Robert A. Sarr. The Art of Framing:Managing the Language of Leadership. San Francisco: Jossey-Bass.

Fine, G.A., and S. Kleinman. 1979. Rethinking subculture: an interactionist analysis. *American Journal of Sociology* 85( 1): 1-20.

Fisher, R., and W. Ury.. 1981. Getting to Yes. Middlesex, England: Penguin Books.

Flanagan, Patrick. 1995. The ABCs of changing corporate culture. *Management Review* (July): 57-61.

Forsythe, G. B. 1992. The preparation of strategic leaders. Parameters (Spring): 38-49.

Franks, F. Guilford, S.G. Isaksen, and D. J. Treffinger. 1985. Creative Problem Solving: *The Basic Course. Buffalo,* NY: Bearly Limited.

Gabriel, R.A. 1985 Military Incompetence: Why the Military Doesn't Win. New York: The Noonday Press.

George, Alexander L., Presidential Decisionmaking in Foreign Policy: The Effective Use of Information and Advise. Boulder: Westview Press, 1980.

Goldberg, R. A. 1997. Talking about change. Issues and Observations 17. Greensboro, NC: The Center for Creative Leadership.

Goodfellow, B. 1985. The evolution and management of change in large organizations. *Army Organizational Effectiveness Journal* 1: 25-29.

Graham, Bradley. 1997. Washington Post, May 11, A 19.

Gran, L. 1997 Monsanto's bet: There's gold in going green. Fortune, April 14, 116-118.

Hackman, J.R., ed. 1990. Groups That Work And Those That Don't. San Francisco: Jossey-Bass.

Hallowell, R. 1996 Southwest Airlines: A case study linking employee needs satisfaction and organizational capabilities to competitive advantage. *Human Resource Management* 35( 4): 519-521.

Hambrick, D.C. 1995 Problem CEOs have with their top management teams. *California Management Review* 37( 3): 110-127.

Harari, O. 1994. Beyond the "vision thing." Management Review (November): 29-31.

Harari, Oren. 1995. Three vital little words. Management Review (November): 25-27.

Harvey, J. B. 1988. The Abilene Paradox and Other Meditations on Management. Lexington, MA: Lexington Books.

Hillerman, Tony. 1988. A Thief of Time. NY: Harper.

Hogan, R., G. J. Curphy, and J. Hogan. 1994. What we know about leadership. *American Psychologist* 49: 493-504.

Huber, G.P. 1991 Organizational learning: The contributing processes and the literatures. *Organization Science* 2(1): 88-115.

Hunt, J. G. 1991. Leadership: A New Synthesis. Newbury Park, CA: Sage Publications.

Jacobs, T. O., and E. Jaques. 1990. Military executive leadership. In K.E. Clark and M. B. Clark, Measures of Leadership. West Orange, NJ: Leadership Library of America, Inc.

Jacobs, T. O. 1996. A Guide to the Strategic Leader Development inventory. Washington, D C: National Defense University.

Jacobs, T.O. 1996. Strategic pitfalls and opportunities: power and politics. In A Course Text Strategic Decision Making. Washington, DC: Industrial College of the Armed Forces, National Defense University,

Jacobs, T.O., and E. Jaques. 1987. Leadership in complex organizations. In Human Productivity and Enhancement. Edited by J. A. Zeidner. Organization and Personnel, vol. 2. New York: Praeger.

Jacobs, T. O., and E. Jaques. 1991. Executive leadership. In Handbook of military psychology. Edited by R. Gal and A. D. Manglesdorff. Chichester, England: Wiley.

Jacobs, T. O. undated. Cognitive behavior and information processing under conditions of uncertainty. Alexandria, VA: U. S. Army Research Institute for the Behavioral and Social Sciences.

Jacobs, T.O. 1996. A Guide to the Strategic Leader Development Inventory. Washington, DC.: Industrial College of the Armed Forces, National Defense University.

Janis, I. L. 1983. Groupthink. In vol 2, Small groups and social interaction. Edited by H. H. Blumberg, A. P. Hare, V. Kent, and M. F. Davis. New York: Wiley.

Jaques, E., and S.D. Clement. 1991. Executive Leadership: A Practical Guide to Managing Complexity. Arlington, VA: Cason Hall.

Jaques, E. 1986. The development of intellectual capability: A discussion of Stratified Systems Theory. *Journal of Applied Behavioral Science* 22, 361-384.

Jaques, E. 1989. Requisite organization. Arlington, VA: Cason Hall.

Jefferies, C.L. 1992. Defense decision making in the organizational-bureaucratic context. In American Defense Policy. Edited by J. S. Endicott and Roy W. Stafford, Jr. 4th ed. Baltimore MD: The Johns Hopkins Unversity Press.

Johns, John H. 1988. Ethical dimensions of national security. In Bureaucratic Politics and National Security: Theory and Practice. Edited by David C. Kozac and James M. Keagle. Boulder, CO: Lynne Rienner Pub.

Katz, D., and R. L. Kahn. 1978. The Social Psychology of Organizations. 2nd. ed.. New York: Wiley.

Katzenbach, J.R. 1993 The Wisdom of Teams: Creating the High Performance Organization. Boston: Harvard Business School Press.

Katzenbach, Jon R., and Douglas K. Smith. 1993. The discipline of teams. *Harvard Business Review* (March-April):111-120.

Kenny, D. A., and S. J. Zaccaro. 1983. An estimate of variance due to traits in leadership. *Journal of Applied Psychology* 68: 679-685.

Kirton, M., Laskey, V., and T. E. Lawrence. 1929. The science of guerilla warfare. *Encyclopedia Britannica.*

Knowlton, B., and M. McGee. 1994. Strategic Leadership and Personality: Making the MBTI® Relevant. Washington, DC: National Defense University.

Kotter, J.P. 1985. Power and Influence: Beyond Formal Authority. New York: Free Press.

Kotter, J.P. 1978. Power, success and organizational effectiveness. *Organizational Dynamics* 6 (3): 27-40.

Kotter, J. P. 1995. Leading change: Why transformation efforts fail. *Harvard Business Review* (March-April): 59-67.

Krepinevich, Andrew F. 1995. Restructuring for a new era: framing the roles and missions debate. Defense Budget Project, Washington, DC.

Larson, C.E., and F.M.J. LaFasto. 1989 Teamwork: What Must Go Right, What Can Go Wrong. Newbury Park: SAGE.

Larwood, L., C.M. Falbe, M.P. Kroger, and P. Miesing. 1995. Structure and meaning of organizational vision. *Academy of Management Journal* 38( 3): 740-769.

Lasswell, H.D. 1936. Who Gets What, When, How. New York: McGraw-Hill.

Lawrence, P., and Lorsch, J. 1967. Organization and environment. Boston, MA: Harvard Business School Division of Research.

Lawrence, P. 1968. How to deal with resistance to change. *Harvard Business Review* (January-February).

Lax, David A., and James K. Sebenius. 1986. The Manager as Negotiator: Bargaining for Cooperation and Competitive Gain. New York: The Free Press.

Levinson, Daniel J. 1978. The Seasons of a Man's Life. New York: Alfred A. Knopf.

Lewicki, Roy J., and, Joseph A. Literer. 1985. Negotiation. Homewood, IL: Richard D. Irwin, Inc.

Lewis, P. M., and T.O. Jacobs1992. Individual differences in strategic leadership capacity: A constructive/developmental view. In Strategic Leadership: A Multiorganizational Perspective. Edited by R. L. Philips and J. G. Hunt. Westport, CT: Quorum Books.

Litterer, Joseph A. 1973. The Analysis of Organizations. John Wiley and Sons, Inc. New York.

Louis, M.R. 1980. Organizations as culture-bearing milieux. In Organizational Symbolism. Edited by L.R. Pondy, et al. Greenwich, CT: JAI.

Louis, M.R., B.Z. Posner, and G.N. Powell. 1983. The availability of socialization practices. *Personnel Psychology* 36( 4): 857-866.

Lucas, K. W., and J. Markessini. 1993. Senior leadership in a changing world order: Requisite skills for U. S. Army one- and two-star generals. ARI Technical Report No. 976. Alexandria, VA: U. S. Army Research Institute for the Behavioral and Social Sciences.

Luft, J. 1984. Group Processes: An Introduction to Group Dynamics. 3rd ed. Palo Alto, CA: Mayfield.

Machiavelli, N. 1513. The Prince. Middlesex, England: Peguin Books, Ltd.

MANDATA, Inc. 1994. MANSPEC Counsellor's Guide.

Mapes, James. Foresight first. 1991. Sky (September): 96-98.

Martin, J., and C. Siehl. 1983. Organizational culture and counterculture: an uneasy symbiosis. *Organizational Dynamics* 12 ( 2): 52-64.

McCall, Morgan, Michael Lombardo, and Ann Morrison. 1988. The Lessons of Experience: How Successful Executives Develop on the Job. Lexington, Ma: Lexington Books.

McCall, M. W., and M. M.Lombardo. 1983. Off the Track: Why and How Successful Executives Get Derailed. Englewood Cliffs, NJ: Prentice Hall.

McCauley, M. H. 1990. The Myers-Briggs Type Indicator and Leadership. In K. E. Clark and M. B. Clark, Measures of Leadership. West Orange, NJ: Leadership Library of America, Inc.

McCrae, R. R., and P. T. Costa, Jr. 1987. Validation of the five-factor model of personality across instruments and observers. *Journal of Personality and Social Psychology* 52: 81-90.

McGee, M. L. 1997. MBTI® correlates of strategic leader potential: Initial support for a misunderstood theory. In Proceedings of the Second Annual International Research Conference on Leadership and the Myers-Briggs Type Indicator®, April 2-4, 1997, Wyndom Bristol Hotel, Washington, DC.

McGee, LTC Mike. 1993. Measuring Up: A Systemic Technology for Developing Leaders. U.S. Army Research Fellow Report. Washington, DC: National Defense University Press.

McGee, M. L., T. O. Jacobs, R. N. Kilcullen, and H. Barber. 1996. Conceptual capacity as competitive edge: Developing leadership for the new Army. Paper presented at the U. S. Army Leadership Symposium, March 27-29, 1996, Chicago, IL: Cantigni Conference Center.

McIntyre, R. M., P. Jordan, C. Mergen, L. Hamill, and T. O. Jacobs. 1993. The Construct Validity of the CPA: Report on Three Investigations. Alexandria, VA: U. S. Army Research Institute for the Behavioral and Social Sciences.

McNamara, Robert 1995. In Retrospect: The Tragedy and Lessons of Vietnam. New York: Random House.

Miller, James. Living Systems. 1992. University Press of Colorado.

Mintzberg, H. 1975. The manager's job: Folklore and fact. *Harvard Business Review* 53: 49-61.

Mintzberg, H. 1973. The nature of managerial work. New York: Harper & Row.

Moore, W. Edgar, Hugh McCann, and Janet McCann. 1985. Creative and Critical ThinkinG. New York: Houghton Mifflin.

Mumford, M. D., S. J. Zaccaro, F. D. Harding, E. A. Fleishman, and R. Reiter-Palmon. 1993. Cognitive and Temperament Predictors of Executive Ability: Principles for Developing Leadership Capacity. Technical Report 977. Alexandria, VA: U. S. Army Research Institute for the Behavioral and Social Sciences.

Mumford, M. D., and M. S. Connelly .1991. Leaders as creators: Leader performance and problem-solving in ill-defined domains. *Leadership Quarterly* 2: 289-316.

Nanus, Burt. Visionary Leadership: Creating a Compelling Sense of Direction For Your Organization. San Francisco, CA: Jossey-Bass Publishers, 1992.

Nixon, R. M. 1982. Leaders. New York: Warner Books.

O'Brien, Bill. 1996. Transforming the character of a corporation. The Systems Thinker 7 (2): 1-5.

O'Reilly, Charles 1989. Corporations, culture, and commitment: Motivation and social control in organizations. *California Management Review* 31: 9-25.

Ouchi, W. 1980. Markets, bureaucracies and clans. *Administrative Science Quarterly* 25: 129-141.

O'Reilly, Charles. 1989. Corporations, culture, and commitment: motivation and social control in organizations. *California Management Review* 31: 9-25.

O'Toole, J. 1995. Leading Change. San Francisco, CA: Jossey-Bass Publishers.

Patton, G. S., Jr. 1947. War As I Knew It. Boston: Houghton Mifflin Company.

Perez-Reverte, Arturo. 1994. The Flanders Panel. New York: Harcourt Brace.

Peters, Tom. 1987. Thriving on Chaos. NY: Alfred A. Knopf.

Pfeffer, J. 1981. Power in Organizations. Marshfield MA: Pitman Publishers.

Pfeffer, J. 1992. Managing With Power: Power and Influence in Organizations. Boston MA: Harvard Business School Press.

Priem, Richard L. 1990. Top management team group factors, consensus, and firm performance. *Strategic Management Journal* 11: 471-476.

Quinn, R. E. 1988. Beyond rational management: Mastering paradoxes and competing demands of high performance. San Francisco: Jossey Bass.

Russo, Edward J., and J. H. Shoemaker. 1989. Decision Traps. New York: Doubleday.

Sackmann, S.A. 1992. Culture and subcultures: an analysis of organizational knowledge. *Administrative Science Quarterly* 37: 140-161.

Salancik, C. R., and J. Pfeffer. 1977. Who gets power—and how they hold on to it:a strategic contingency model of power. Organizational Dynamics 5: 3-21.

Schein, E. 1990. Organizational culture. *American Psychologist* 45 (2): 109-119.

Schein, E. H. 1988.Organizational Culture and Leadership. San Francisco: Jossey-Bass.

Scholtes, Peter R. 1988. The Team Handbook How to Use Teams to Improve Quality. Madison, WI: Joiner Associates Inc.

Schweiger,David M., andWilliam R. Sandberg ,William R. 1989. The utilization of individual capabilities in group approaches to strategic decision-making. *Strategic Management Journal* 10: 31-43.

Senge, P. M. 1990. The Fifth Discipline: The Art and Practice of the Learning Organization. New York: Doubleday.

Sheehy, Gail. 1995. New Passages: Mapping Your Life Across Time. New York: Random House.

Smith, Hedrick. 1988. The Power Game. New York: Ballantine Books.

Stamp, G. P. 1978. Assessment of individual capacity. In Levels of Abstraction in Logic and Human Action. Edited by E. Jaques, R. O. Gibson, and D. J. Isac. London: Heinemann.

Stamp. G. 1988. Longitudinal Research into Methods of Assessing Managerial potential, Technical Report 819. Alexandria, VA: U. S. Army Research Institute for the Behavioral and Social Sciences.

Starbuck, W.H., and F. J. Milliken, F.J. 1988 Executive perceptual filters filters: what they notice and how they make sense. In The Executive Effect: Concepts and Methods for Stuidying Top Managers. Edited by D.C. Hambrick. Greenwich: JAI Press.

Stewart, S. R., R. Archer, H. Barber, P. Tuddenham, and T. O. Jacobs, 1993. An Instructional Technology for Developing Metacognition and Creative Thinking Skills. Alexandria, VA: U. S. Army Research Institute for the Behavioral and Social Sciences.

Streufert, S., and R. W. Swezey. 1986. Complexity, managers, and organizations. Orlando, FL: Academic Press.

Streufert, S., and S. C. Streufert. 1978. Behavior in the complex environment. New York: John Wiley.

Sullivan, G. R., and M. V. Harper. 1996. Hope is Not a Method. New York: Random House.

Sulsky, Lorne M., and David V. Day. 1992. Frame o-of-reference training and cognitive categorization: an empirical investigation of rater memory issues. *Journal of Applied Psychology* 77 (4): 501-510.

Thayer, L. 1988. Leadersip/communication: A critical review and a modest proposal. In Handbook of Organizational Communication, eds. G.M. Goldhaber and G.A.Barnett. Norwood, NJ.:Ablex.

Strategic Leader Development Inventory. 1993. Alexandria, VA: The Army Research Institute for the Behavioral and Social Sciences.

Theobald, R. 1994. Visionary planning for a compassionate era. *Planning Review* 225: 12-14.

Thurman, M. 1991. Strategic leadership. In Strategic Leadership Conference Proceedings.. Carlisle, PA: U. S. Army War College and U. S. Army Research Institute.

Trice, H.M., and J. M. Beyer, J.M. 1984. Studying organizational cultures through rites and ceremonials. *Academy of Management Review* 9: 653-669.

Trice, H.M. Rites and Ceremonials in Organizational Culture. 1988. In Vol. 4., Perspectives on Organizational Sociology: Theory and Research. Edited by S.B. Bacharach and S.M. Mitchell. Greenwich, CT: JAI Press.

Trice, H.M., and J. M. Beyer. 1993. Cultures of Work Organizations. Englewood Cliffs: Prentice Hall.

Tsouras, P. G. 1992. Warriors' Words: A Quotation Book. London: Arms and Armour Press.

Tsui, A. S. 1984. A multiple constituency framework of managerial reputational effectiveness. In Leaders and managers: International perspectives on managerial behavior and leadership. Edited by J. G. Hunt, D. Hosking, C. Schriesheim, and R. Stewart. New York: Pergammon Press.

Ury, William. 1993. Getting Past No: Negotiating Your Way From Confrontation to Cooperation. Rev. ed. New York, NY: Bantam Books.

Vancouver, J.B. 1996 Living systems theory as a paradigm for organizational behavior: understanding humans, organizations, and social processes. *Behavioral Science* 41( 3): 165-204.

VanMaanan, J. and S. R. Barley. 1984. Occupational communities: culture and control in organizations. *Research in Organizational Behavior* 6: 267-365.

Walsh, J.P. 1988. Selectivity and selective perception: an investigation of managers' belief structures and information processing. Academy of Management Journal 31 (4): 873-896.

Walton, R.E., and J. R. Hackman. 1986. Groups under contrasting management strategies. In Designing Effective Groups. Edited by Paul S. Goodman and Asso. San Francisco: Jossey-Bass.

Weick, Karl E. Sensemaking in organizations. 1995. Fouindations for Organizastional Science. Thousand Oaks, CA: Sage Publications, Inc.

Weisbord, Marvin R. 1992. Discovering Common Ground. San Franciso: Berrett-Koehler Publishers.

Wilgoren, Debbi, The Washington Post, March 28, A19.

Willbern, York. 1984. Types and Levels of public morality. Public Administration Review (March-April): 102-108.

Wuthnow, R., and M. Witten, M. 1988. New directions in the study of culture. *Annual Review of Sociology* 14: 50-51

Yates, Douyglas Jr. 1987. The Politics of Management .San Francisco: Jossey-Bass Publishers.

Yukl, G. 1994. Leadership in Organizations. Englewood Cliffs, NJ: Prentice Hall.

Zaccaro, S. J. 1996. Models and Theories of Executive Leadership: A Conceptual/ Empirical Review and Integration. Alexandria, VA: U.S. Army Research Institute for the Behavioral and Social Sciences.

Zald, M.N., and M.A. Berger. Social movements in organizations. *American Journal of Sociology* 83 (4): 240-259.

Zsambok, Caroline E., Gary Klein,, Molly M. Kyne, and David W. Klinger. 1992. Advanced Team Decision Making: A Developmental Model. Fairborn, OH: Klein Associates Inc.

www.ingramcontent.com/pod-product-compliance
Ingram Content Group UK Ltd.
Pitfield, Milton Keynes, MK11 3LW, UK
UKHW021948270726
14060UKWH00002B/426